"I love how Bill Best 'stirs the pot.' Going to his house and sitting at his table after a walk through the garden will reveal the best-tasting tomatoes and, likely, some Turkey Craw beans—my personal favorite. But Bill also stirs the pot metaphorically by showing the Appalachian region and the world how place matters in a transnational political economy that has long said otherwise. For all the talk and attention given to globalization, Bill Best in his life's work and especially in this delightful book proves that place matters. The local is the place of deep, abiding, but also fragile knowledge. If you doubt it, ask your heart and your tongue. They know."

—Chad Berry, Goode Professor of Appalachian Studies and Professor of History at Berea College, author of *Southern Migrants, Northern Exiles.*

"The magic in the greatest of all Jack tales is that what appears to be a mere handful of seeds turns instead into a giant beanstalk leading to riches beyond measure. That same sort of alchemy is at work here in Bill Best's *Saving Seeds, Preserving Taste.* Yes, it's a practical and useful handbook for good garden husbandry, but as it unfolds before your eyes, it reveals as well a vital world of southern Appalachian people, plants, food, and practice to nourish both body and soul."

—Ronni Lundy, founding member of the Southern Foodways Alliance, author of *Shuck Beans, Stack Cakes, and Honest Fried Chicken*

"It was a simple packet of beans purchased from Bill Best that restored my restless spirit. Last winter, we finally met. I could not tell him how I felt because I knew I would cry. But watching Bill Best instruct the youthful Chef Jeremy Ashby in the finer points of heirloom seed saving and history, I found my heart was filled with boundless joy. The legacy will continue! That one moment was worth the trip up the Mountain Parkway. God bless you, Bill Best . . . for you have blessed me and the people of Appalachia for generations to come. May your harvest always be plentiful and the bean beetles few!"

—Joyce Pinson, *Edible Ohio Valley*

Saving Seeds, Preserving Taste

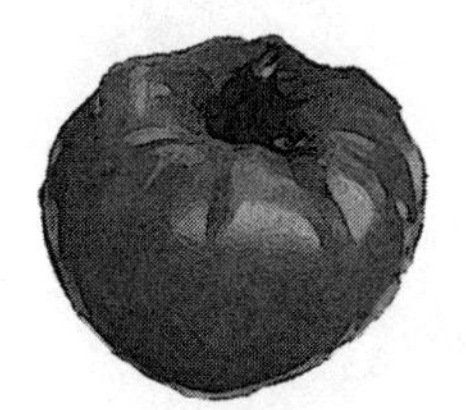

heirloom seed savers in appalachia

BILL BEST

Foreword by Howard L. Sacks

Ohio University Press

Athens

Ohio University Press, Athens, Ohio 45701
ohioswallow.com

Printed in the United States of America
Ohio University Press books are printed on acid-free paper ∞ ™

23 22 21 20 19 18 17 16 15 14 13   5 4 3 2 1

Library of Congress Cataloging-in-Publication Data
Best, Billy F., 1935–
Saving seeds, preserving taste : heirloom seed savers in Appalachia / Bill Best.
p. cm.
Includes bibliographical references and index.
ISBN 978-0-8214-2049-2 (pb : alk. paper) — ISBN 978-0-8214-4462-7 (electronic)
1. Seeds—Harvesting—Appalachian Region, Southern. 2. Vegetables—Seeds—Appalachian Region, Southern. 3. Vegetables—Heirloom varieties—Appalachian Region, Southern. 4. Fruit—Seeds—Appalachian Region, Southern. 5. Fruit—Heirloom varieties—Appalachian Region, Southern. 6. Seed exchanges—Appalachian Region, Southern. I. Title. II. Title: Heirloom seed savers in Appalachia.
SB118.38.B47 2013
631.5'2097569—dc23

2012050529

Contents

Foreword

Howard L. Sacks

A COLLEAGUE OF mine recently shared an eyebrow-raising story involving her college course "Botany and Botanical Arts." In an "unknown plant" assignment, students were given different "mystery" seeds that they were challenged to cultivate, observe, and identify. One day my colleague discovered a student in the greenhouse who was at a loss over how much to water the plants; the exasperated student explained that she had never before planted a seed.

How removed we have become from an act so fundamental to human civilization! Planting a seed—horticulture—prompted our early ancestors to abandon a nomadic life of foraging to take up a more sedentary existence. The newfound dependability of the food supply enabled populations to grow. Individuals could accumulate more possessions, because they were no longer required to carry everything with them from one food source to the next. Differences of wealth emerged, and with that, differences of power. Humans grew increasingly territorial, and hostilities became more commonplace as groups sought to protect their cropland. All this, from the planting of a seed.

Not so long ago, most Americans still planted a few seeds. I grew up in postwar Philadelphia during the first wave of the new American idyll known as suburbia. On my street, partially prefabricated identical homes were perched like so many Monopoly plastic houses on deforested land. Visitors to our place were directed to "the sixth new house on the right." The joke was that a drunken husband might walk into the wrong house at night.

But my mother had grown up in the mountains of Pennsylvania, where houses had been hand built of brick and stone and wood, and the countryside had not seemed far removed from town life. Yet housewives like her and their businessman or professional husbands could not sign up quickly enough for this spot of perfection bearing the fatuous title of "Westgate Hills" (no gates, no hills, but indeed west of the city). Something, though, pulled at my mother, because from my earliest memory she had always planted a few tomatoes against the foundation of our split-level house so that we could taste something fresh. *Fresh*—the word itself is radical, given the change in the American diet to canned and frozen foods. Peering down from my bedroom window to see my neighbor scratching the dirt in his own makeshift backyard garden, and paying attention to my mother and her fondness for these optimistic young tomato plants, I saw that things could live and grow, in opposition to a profoundly denatured landscape.

Then, as now, we knew no more about the source of the seeds we planted than about the origins of the food we purchased from the supermarket shelves. Both seeds and foods were identifiable to us by their corporate names, whether Burpee, Gerber, or Heinz. It wasn't always that way, of course. Our collective ignorance can be traced to the mid-twentieth-century revolution in agriculture that transformed a diffuse, regionally based system of growing food to a highly centralized system of commodity production for a global market. By the late 1940s, tractors and combines had largely replaced machinery drawn by mules and horses, enabling farmers to cultivate more land with less reliance on their neighbors at planting and harvest time. In the following decades, chemical fertilizers, herbicides, and pesticides were promoted and adopted as the saviors for increased productivity and crop yield—their environmental costs left unquestioned in this campaign directed toward farmers. Most recently, genetic seed modification has enabled new varieties of fruits and vegetables designed specifically for global transport and marketing.

These technological breakthroughs complemented government policy and corporate interests. In the 1970s, U.S. Secretary of Agriculture Earl Butz admonished farmers to "get big or get out," and low-interest bank loans enabled many farmers to buy more land and new equipment to produce a single commodity for sale on the global market. Agriculture schools at land-grant universities spearheaded research to develop new chemical and biological innovations, often financed by the very companies that would reap the profits as farmers became increasingly reliant on their products. Promises of an ever-expanding global demand for American agricultural commodities fueled agricultural speculation, and for a short time many American farmers did just fine. Tragically, the bubble eventually burst, and the reverberations have been both devastating and long-lasting. In the short term it meant the wholesale loss of farms and the decline of rural communities, and the farm crisis anticipated the recent housing market collapse that severely battered the American economy.

In the late twentieth century, American consumers were invited to understand modern agriculture as an unmitigated good. On television, in magazines, in the school systems, and at the university ag program level, the discourse was nearly exclusively that our food supply was abundant, affordable, and convenient. But a growing number of families have begun to question the wisdom and sustainability of handing over our dinner plates to industrial farming enterprises. People are again asking questions about the sources of their food. For some, the issue is health—childhood diabetes, adolescent eating disorders, heart disease, and other diet-related concerns. Others bring agricultural practice under closer scrutiny over food safety, as *E. coli*, *salmonella*, and *mad cow disease* enter our everyday lexicon. And concern about fossil fuels focuses the lens on the degree to which our food supply depends on oil: gas for the combine, petroleum-based fertilizers, and the cost of long-distance transportation, with their corresponding impact on food prices.

And then there is the matter of taste. Tomato varieties designed for resistance to bruising during cross-country transport and extended shelf life just do not taste very good, particularly when compared with homegrown varieties. Eggs from chickens raised under megafarm confinement conditions often have about as much taste as the cardboard containers in which they are packaged. And while supermarkets contain an astonishing array of products, the apples or greens in those bins actually represent just a few varieties chosen for their consistent, blemish-free appearance. As the song goes, "All made out of ticky tacky, and they all look just the same." When you take the time to think about it, it is no coincidence that suburbanized uniformity shows up in our food system.

But consider an alternative world, one in which it is not only professionals who dictate the colors and textures on our plates, and in which memory and local culture go into every child's lunchbox. This is the world of the seed savers of the Southern Appalachians. From the men and women who practice seed conservation as part of daily life, we can learn important lessons about eating wholesomely and living more holistically.

Until the advent of commercially available seed stock, the practice of saving harvested seeds for future planting was born of necessity; sustainability of the food supply was an immediate, ever-present concern. The European pioneers who settled in western North Carolina and Kentucky brought seeds from their homelands and the eastern colonies, obtaining others from the indigenous Native people who had long planted varieties of corn.

The varieties that flourished in the upper South were carefully preserved, and new strains that proved desirable—created through natural mutation or deliberately, by crossbreeding experimentation—were added to the local seed stock. This approach has yielded a diverse array of beans, corn, tomatoes, apples, and other fruits and vegetables prized for their flavor, texture, productivity, suitability for preserving, and eye appeal. In short, they both taste good and work well in the locale.

This localized method of creating, disseminating, and preserving varieties stands in sharp contrast to the commercial system of genetic modification, corporate patents, and global marketing. As the stories in this volume reveal, the preservation of a particular variety of bean can often be traced to the dedicated efforts of a single individual over many years, experimenting with a new variety or simply planting, harvesting, and preserving seeds to ensure that a longtime favorite makes it into succeeding generations. Seed sharing follows the contours of traditional community life, as gardeners distribute a variety to family members, friends, and neighbors. Some growers freely share a prized bean with anyone likely to cultivate it, in order to ensure its preservation. Others more proprietary by nature might hoard their stock, prompting a neighbor to raid a garden late at night in an act of seed liberation. Even today some rural communities continue the barter system of acquiring goods and services, and seeds play a role in such exchanges: they may be traded for supplies, slipped across the table at a church picnic, or offered to entice support for a political candidate. Exchanging seeds clearly produces more than food; it is an act of profound social meaning, nurturing community and family bonds.

At the hardware store, the church supper, or the family reunion, telling stories about seeds and the giving of seeds constitute a distinct type of knowledge exchange. This system of knowledge creation and transmission challenges the dominant narrative about who is an authority and whose knowledge prevails in society. Unlike academic and corporate professionals, who tend to speak mainly to their peers in journal articles and who see seeds as commodities for patent protection, the experts in the world of seed saving are imbedded in their communities, and they have built their knowledge base—and, frankly, their passion—through a lifetime of in-the-field experience and careful observation. The displacement of just this sort of local knowledge is what marked the transition to modern agriculture, with the advent of university-trained agriculture experts selling their version of a brighter future at rural farmers' institutes and extension offices. The speakers in this volume return us to the once-prevalent, surprisingly persistent world of neighbors with brains worth picking.

Author Bill Best brings alive a range of keepers, many with specialized knowledge. We meet, for example, an expert in tree grafting, a critical skill for preserving heirloom fruits. Many varieties of beans are named after the women who developed and preserved them, affirming the primary role of women as knowledge bearers. At the same time, the author demonstrates how the modern seed-saver network actively incorporates the Internet, providing an interesting case study of the interplay of orally transmitted traditional knowledge and modern technology.

The author's stories about gardening convey a deep sense of regional history and folklore. We learn about Daniel Boone's pioneer exploits, the Trail of Tears that removed Cherokees from their homeland, methods of tobacco cultivation, local politics, and the recent migrations that have shaped the transmission of seeds across space and time.

Beyond the obvious functional impact of seed saving, seeds and plants feature significantly in Appalachian expressive culture. As Best notes, one need look no further to appreciate the cultural significance of beans than "Jack and the Bean Stalk" and the other Jack tales, stories famous in western North Carolina and eastern Kentucky, where the author has spent his life. Among the seed savers mentioned is Letha Hicks, drawing us to note the connection of saving seeds and saving stories. Jack tales, of European and Celtic origin, were preserved in America principally by the Hicks family of the Beech Mountain region of western North Carolina. Ray Hicks, who in 1983 received a National Heritage Fellowship from the National Endowment for the Arts as a teller of these tales, learned them as a boy from his elders, who told the stories while canning or drying apples. In a region where beans are a staple of necessity and hardship presents many obstacles, it is easy to understand the appeal of stories in which magical beans and individual pluck enable success.

One of the most popular fiddle tunes of the Southern Appalachians is "Leather Britches," a title that refers to a way of preserving long beans. Before the widespread use of freezing or canning, people would string mature beans together and air-dry them for several weeks, preserving these "leather breeches" (britches) for later rehydration and cooking.

In the broadest sense, seed saving is an act of connection to place. Heirloom varieties bear the names of the people, animals, materials, and motivations that define local life. When we read of Ora's Speckled Bean, Brown Goose, White Case Knife, and Radiator Charlie's Mortgage Lifter, we have a sense of a story behind each one. The community that sustains these heirloom fruits and vegetables stands as a telling counterpoint to the contemporary notion that rapid mobility and separation from friends and family are what life is and should be—the uncritically accepted norm. A poignant reminder of seeds as a connection to place is the effort of those who have left Appalachia to secure and cultivate the varieties they knew back home.

Again in contrast to the big and the uniform in agriculture, Best refocuses our attention to an intimate scale. He invites us to notice the distinctive texture of a greasy bean, the compactness of beans in a pod, or how a tomato seems to taste better when accompanied by the smell of field tobacco.

Today, local efforts to preserve heirloom seeds have become part of a growing national movement. Seed swaps among neighbors at the local store have given way to a network of enthusiasts, regional educators, and nonprofit groups exchanging on the Internet. "Local" is harder to define these days, surely. Yet it is fair to say that these savers constitute an alternative agricultural world, one that operates on assumptions and values that sharply contrast with those of global agribusiness.

There is undeniably an element of the romantic in tending to seeds as if life depends on them. But it is only modern rationality, with its devotion to finding a technical solution to any problem, that prompts us to reject the romantic as superfluous. Perhaps there is inherent value in the dirty fingernails and slightly aching spine, and the curiosity and dedication that people bring to toiling in a small garden and helping plants grow. These stories offer a critical perspective on our own lives, beginning with what we sit down to eat at the dinner table. All this, from the planting of a seed.

Howard L. Sacks teaches sociology at Kenyon College, where he directs the Rural Life Center. For the past fifteen years, he has led an initiative to build a sustainable local food system in Knox County, Ohio. He and his wife, Judy, raise sheep on their farm near Gambier.

Dedication

My mother, Margaret Sanford Best, was an old-time seed saver who took her seed saving very seriously. Born over a hundred years ago in 1911, at the time of her death in 1994, just four weeks before her eighty-third birthday, she was still busy trading seeds with extended family members and other people in the Upper Crabtree community in Haywood County, North Carolina. Having said frequently that she would wear out rather than rust out, Mother had kept gardening as long as possible, always saving seeds for the next season and making sure she had plenty to share.

One of my earliest memories is picking colorful cornfield beans with her and learning how to avoid the equally colorful saddleback caterpillars and other stinging "worms" that could leave painful welts on bare skin. Mother picked the beans higher up the cornstalks, and I picked those close to the bottom. That I had helped pick the beans I ate for supper that night made me feel very much a part of the family life and farm life in which I was a participant. I was about three years old at the time.

I had my first garden of my own in 1963, after my wife and I had started working at Berea College. As soon as we had arrived in Kentucky, we had bought a farm in Jackson County, and I ordered seeds from catalogs, mistakenly assuming that they would be like the beans and tomatoes I had grown up with. I was quite disappointed, to put it mildly. While visiting family in North Carolina that fall, I mentioned my disappointment with commercial beans, and Mother quickly went to her can house and handed me bags of several of her varieties of beans, which were beginning to be called heirlooms. I have never looked back.

Mother realized early that the commercial seed companies had stopped selling the beans that had flavor and texture worthy of the name *bean.* She intensified her seed-saving activities as she got older, seemingly aware that her efforts would be a legacy worth leaving to her descendants and any others who might like to grow and eat these beans.

Ten years after her death, my youngest sister, Janet, who lives with her husband in the house all five children in our family were raised in, asked me to check Mother's freezer, since it appeared that several of her bean containers full of bean seeds were still in the freezer. I discovered thirteen varieties of beans still in her freezer, untouched since the day she died. The following summer I grew some of all thirteen varieties, with all thirteen having good germination.

Seven years later, in 2010, Janet cleaned out the entire freezer and defrosted it, only to find several more packages of beans dating back to 1978. She spilled some of the 1978 beans and discovered three days later that they had germinated in the water left by the melting ice. Most exciting to Mother's five children was to find some cut-short beans that all of us had remembered from our growing up.

For the reasons listed and many others, I am dedicating this book to my mother and hope that I can make a contribution to other people as she made to me and others with her hard farming and gardening work and her sage advice.

Preface

A FEW YEARS ago, savers of heirloom seeds were thought to be a little bit eccentric or worse. After all, everyone knew that the many seed companies peddling their wares were looking out for America's gardeners and maintaining an abundance of varieties for each and every purpose and growing condition.

Gardening fell out of favor with many Americans as "Super" markets made available more selections than most people had ever known. The United States pursued a cheap food policy, with land grant universities leading the charge to make food available as cheaply as possible, and with the government also making surplus foods available to public schools and other agencies.

But somewhere along the food superhighway, there came to be a few bumps in the road. Small seed companies were swallowed up by larger seed companies, and larger seed companies were swallowed up by international food, feed, seed, and chemical conglomerates that tended not to take very seriously their responsibilities in maintaining genetic diversity and producing quality foods.

Food plants were genetically modified to make mechanical harvesting and long-distance shipping over great distances easier. Vegetables were toughened up and made more uniform so that they could be harvested by machine with one pass at one time. American food production left the "Garden State" and other states close to population centers and moved to California and Florida, if not as far afield as Mexico, South America, and even Europe and Asia.

Genetic engineering replaced the preservation of genetic diversity, and companies ridded themselves of thousands of varieties that had been maintained by the smaller companies that were cannibalized during the consolidation process. This even included many of the early hybrids that were bred for flavor, texture, and nutrition. Suddenly toughness was the byword for all things fruit and vegetable.

But a funny thing happened on the way to modernity. Many people started having doubts about the brave new world of genetic engineering, food-borne diseases, childhood obesity, adult-onset diabetes in young children, and food companies using the courts to squeeze out the small farmers by patenting the pollen in the air. Suddenly a collective "Enough!" was heard from sea to shining sea.

This book is about a small part of that "Enough!" We eccentrics are now being heard.

Saving Seeds, Preserving Taste

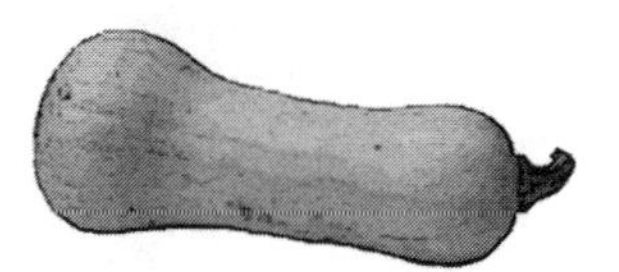

An Introduction to Heritage and Heirloom Seed Saving

I GREW UP believing that the Goose Bean was discovered by my great-grandfather Sanford. My mother had told me that he had shot a wild goose and her grandmother had discovered some bean seeds in its craw as she was dressing it for a meal. The beans were planted, grew to maturity, had a good flavor, and became one of many varieties of beans kept by our family.

Years later I discovered that many children in the Southern Appalachians had been told the same story by their parents. Essentially the same tale was also told about the Turkey Craw Bean: a wild turkey had been shot for food, bean seeds were found in its craw, and the seeds had been planted and found to be among the best beans around.

The Goose Bean is also known as the Goose Craw Bean and in some areas as the Goose Neck Bean. The Turkey Craw Bean is

Margaret Sanford Best at age eighty. *Photo by Paul Toti*

sometimes just called the Turkey Bean. Both beans are among the favorites of thousands of families in the Southern Appalachians and in other parts of the country where many Appalachian families have migrated.

As is true of many other families in our community, beans were very important to us. When we visited my grandmother Sanford most Sunday afternoons, as a very young child I followed her and my mother to Grandma's garden. They talked gardening while I explored and sometimes listened to their conversations. I later realized that Grandma Sanford was continuing to pass on her gardening traditions to Mother, who was later to pass them on to me. And Grandma Sanford was passing on traditions she had learned from her family decades earlier. Perhaps the most important tradition being passed on was seed saving.

What is important here is the fact that there are hundreds, if not thousands, of heirloom bean varieties maintained by gardeners in the Southern Appalachians. They are also preserved, often in their purest forms, by Appalachian migrants to other parts of the country. Many people migrated to places as far away as Washington State and took their beans with them. For example, there is a bean in Washington State that is called the Tarheel Bean, which, I have been told by several people, was taken from the Jackson/Haywood County area of North Carolina. (My mother's oldest, and only, sister migrated with her husband from Haywood County to Kelso, Washington, in 1918 to work in the timber industry.) Another bean variety now in Washington State was sent to me by a retired Forest Service employee who had taken it with him from West Virginia when he retired. And, of course, there is the famous Trail of Tears Bean, taken from western North Carolina and northern Georgia by the Cherokees when they were forced out of the Southern Appalachians into the Indian Territory (present-day Oklahoma) by the federal government in the 1830s.

Until I was in my midtwenties and going to college and graduate school and then serving in the army, I helped my family with its gardens and other crops when I was home during the summers. After starting college I was rarely there for bean plantings, but on those occasions I was conscious that we still planted beans that my mother had saved from previous years. By that time I knew that bean seeds could be bought from farm stores and from seed catalogues as well, but there was no point in doing so. There were so many varieties in the general area that it was pointless to pay good (and scarce) money for seeds.

However, when I was in my late twenties and starting to garden on my own with my young family, I purchased some seeds from commercial sources. My wife and I had bought a farm in Kentucky that had land similar to that of my home in Haywood County, North Carolina, with basically the same growing season. Our land had a garden plot that had been used for generations, and the soil was exceptionally fertile.

That first summer we had a bumper crop of good tomatoes, sweet corn, okra, and potatoes, but I was in for a rude awakening because of the toughness of the beans we were growing. It had not occurred to me that beans might become inedible because of the toughness of their hulls at any stage of their development. I certainly did not think that a lot of time and money had been spent by seed companies and universities to create tough beans that would not break during mechanical harvest.

When we visited my parents that following Thanksgiving, I told my mother about our bean experience, and she promptly gave me some of her seeds, which had not been contaminated by the tough gene being used almost universally in commercial beans by that time. Unfortunately, I neglected to put her seeds in our freezer, and the beans had holes in them from bean weevils by the time I got ready to plant them in our garden the following summer.

That was lesson number two: always keep bean seeds in airtight containers and refrigerated or frozen until time to plant them. The first lesson, of course, was to stay away from commercial beans entirely. Before Mother had refrigeration (which she got about

Bill Best with customers at Lexington Farmers Market. *Photo by Michael Best*

the time I left for college), she had put mothballs or hot peppers in her bean containers; that kept the weevils at bay but also gave the beans a bad smell—and your hands, too, when planting them.

When I started selling heirloom beans at the Lexington Farmers Market in the early 1970s, I charged more per pound for them than did vendors selling commercial machine-harvested beans that they had purchased for resale. Not long afterward, I noticed something about the buying habits of many customers: they would buy several pounds of commercial, stringless beans from other vendors and then buy a pound or two of "full" beans from me.

When I finally asked why this was so, my customers told me that they were buying my beans "to flavor" the ones they had already bought. As I came to know more of my customers, I realized that most of the ones who bought my beans to flavor the others

Dr. Graydon Long and Mrs. Virginia Long, customers for over thirty years

were migrants to Lexington from eastern Kentucky or mountain counties in other states. They liked the cheap prices of the commercial beans but also wanted taste and texture if possible. They also liked having a few "real" beans mixed in with the commercial beans, which consisted almost entirely of hulls.

As vendors selling commercial beans, usually bought from produce terminals, started raising their prices to have a higher profit margin and their prices became closer to those of heirloom beans, many, if not most, of the heirloom bean lovers started buying only heirloom beans when they were available. They decided that quality was worth the price, especially if there was little difference between the prices. As more eastern Kentucky transplants started coming to the market, many of them shared some of their family beans with the vendors who were interested in growing them for the market.

Another grower started selling heirloom beans about the same time I started, and we traded seeds with one another. By working together and making referrals to one another, we soon had a lot of customers buying heirloom beans—and not just those people who had grown and eaten such beans in their earlier years. Soon heirloom beans came to be in demand by people who had never eaten any beans other than the commercial, machine-harvested ones.

In 1988, freelance writer Judy Sizemore wrote an article about our small farming operation for *The Rural Kentuckian* (now *Kentucky Living*). In the article, she described the ways we had come to participate in the Lexington and Berea farmers' markets and the fruits and vegetables we had been growing and selling for many years. She also described the heirloom beans and tomatoes we were growing, especially the greasy beans (so named because they have slick hulls and look as if they have a thin coat of grease on them).

Within a week after the article came out, I started getting phone calls and letters from people interested in purchasing some of our greasy beans. Almost without exception the phone calls and letters spoke of the superiority of heirloom beans when compared to commercial beans. Many customers wanted to purchase greasy beans to grow in their own gardens, while several proposed trading seeds with me. Others simply wished to send me beans that had been in their families for generations that they would like to see spread around.

One man from London, Kentucky, came by our house and brought an heirloom bean from Morgan County, Kentucky, called the Nickell Bean. I gave him some of our beans in return but unfortunately did not get his name and address, since I had not yet started formally collecting bean varieties and was unaware of the importance of documentation.

Over the next few months, things began to explode, because people were sending the magazine to friends and relatives within

Brian Best, Ron Robinson, and Kathryn Wallace

Kentucky and in other states. I received eighty-six letters from six states within six months. From those letters and the contacts involved, I became a collector, grower, and distributor of heirloom beans. But I was still working in a very informal way and not documenting enough. I certainly became aware that there were a lot more beans out there than I had thought.

Finally it dawned on me that I was on to something that would come to occupy a lot of my time and energy. I had been slow to realize that I was involved in an activity that dealt with a lot of history and culture and also tapped into widespread unhappiness with the state of the modern food supply—a food supply increasingly dominated by large corporate farms and multinational food/feed/seed/chemical conglomerates. (And this was long before the outbreaks of mad cow disease and the more recent problems with tainted food from other countries and the problems with *E. coli*, salmonella, listeria, and other food-borne pathogens.)

I needed assistance from as many interested people as possible, so in cooperation with my youngest son, an agricultural economics professor, and several other people interested in heirloom fruits and vegetables and sustainable agriculture in general, we formed a not-for-profit corporation, the Sustainable Mountain Agriculture Center, to develop a seed bank that would house the growing number of heirloom beans (and other vegetables) in my collection.

My son, Michael Best, directed the organization for its first three years and then went back to university teaching. I have directed it since that time, taking the position some years after my retirement from Berea College. Other members of the board come from three states and include retired college professors, a graduate school dean, an educational TV producer, and growers of heirloom fruits and vegetables.

When we set up our website, www.heirlooms.org, the organization became the focal point for many seed savers and others wishing to become involved with heirloom gardening. I started receiving Appalachian heirloom beans from people in many states and requests from hundreds of people.

Other growers also bought into the idea of growing and selling quality beans, and today heirloom beans could easily and quickly corner the market if enough were available. But because seeds must be saved and the beans must be picked by hand, there often is not enough supply to meet demand, even with prices at $3.00 to $4.00 or more per pound.

PART 1

HERITAGE FRUIT AND HEIRLOOM SEEDS

Not kudzu!—heirloom cornfield beans in full glory

Beans

Beans occupy an almost mythical status not only in the Southern Appalachians but also in other bean-growing parts of the world. They have been found in Indian burial mounds and in pyramids. When kept in airtight jars, they have been found to be viable after hundreds and even thousands of years.

Corn, of course, also occupies high status in many cultures, with corn and beans being the dominant foods. Together, and accompanied by pumpkins, they occupied a special status within many Indian tribes, who saw them as the Three Sisters, for their growth habits were symbiotic, with the cornstalks providing support for the beans; the beans providing nitrogen for the corn, winter squash, and pumpkins; and the squash and pumpkins providing ground cover to help control weeds.

Tyler Hess (*left*) and Brian Best (*right*) hand-planting bean seeds

Our Appalachian Beans Came from Where?

There are many ongoing discussions about where the beans of the Southern Appalachians originated. The debate has intensified with the coming of the Internet and numerous gardening forums where people swap information, sometimes misinformation, and questions. Seed-saving organizations also have sessions at their annual meetings and informal get-togethers.

I have come to believe that most of the beans found in the mountainous areas of North Carolina, South Carolina, Tennessee, Virginia, West Virginia, Ohio, Kentucky, Georgia, and Alabama can be traced to the Indians living in the area when Europeans first arrived from the flat and coastal areas of what would become the United States of America. Given the bean's tendency to cross and mutate, most of the varieties that now exist in the mountains could have come from a far smaller suite of original beans than one might think.

At the annual meeting of the Kentucky Vegetable Grower's Association in January 2005, the keynote speaker, Dr. Gwynn Henderson of the Kentucky Archeological Survey, gave a talk on the history and possible origins of many of the edible plants of the Southern Appalachians. During her presentation, she showed slides of beans taken from an Indian dump in Jessamine County, Kentucky, that had been discovered during a construction project. Carbon-dated at more than 1,000 years old, they were clearly cut-shorts, one of the dominant types of beans in the Southern Appalachians (called cut-shorts because the seeds are so crowded in the pods that they square off on the ends). Beans from another site in Mason County, Kentucky, were more than 1,400 years old.

Cut-short beans take the shapes of squares, rectangles, parallelograms, trapezoids, and even triangles. Because of the high ratio of seed to hull, they are much higher in protein than other beans are and could have been prized by the Indians for that reason alone. They were also valued by the European settlers; today, they are still treasured by traditional gardeners and, increasingly, by farmers' market customers.

Although the same beans have been in many mountain families for generations, determining where particular varieties originated is difficult. But one may safely assume that where certain varieties predominate, that says something about their development, if not their origin.

I was at a conference a few years ago, and during a discussion of traditional foods, someone mentioned a bean that had been in his family for generations; he still maintained that variety even

Brian Best (*left*) and Tyler Hess (*right*) in bean patch

though he lived in a city far removed from the rural area where we were meeting (on property that had belonged to his great-grandfather). Others present entered the conversation with stories about beans that had been in their families for generations as well, and this led to a discussion about family beans being a part of Appalachian culture, perhaps more so than in other regions, or at least longer.

Family Beans

I know of one bean that can be traced back to before the American Revolution. One of my college classmates, Don Fox from Madison County, North Carolina, was calling one of our mutual friends a few years ago. Intending to dial the number of the friend, he mistakenly called me instead. I took the opportunity to discuss his family bean, which he had shared with me a few years before, brought to me by our mutual friend, Ben Culbertson. (This is one of the ways Appalachian beans get around.)

Don's ancestor, by the name of Banks, had migrated from Scotland to the colonies just as the Revolutionary War was beginning. He fought on the British side and, because he had been on the losing side, found himself in a quandary at the end of the war: he could go back to Scotland or into the mountains of what would be western North Carolina. He chose the mountains and ended up marrying a Cherokee woman. One of her contributions to the marriage was a greasy bean, and that bean has thus been in the Fox family since the early 1780s.

One variety has been in my family for at least 150 years, and within a few miles of where I was raised, there are beans that have been in other families for generations. This tradition of family beans marks one of the important ways Appalachian beans have been preserved and developed.

A lifetime of experience suggests to me that many, if not most, of such family beans came about by mutant beans, usually called "sports," showing up in bean patches. On numerous occasions I have been told of a particular bean showing up in someone's garden, usually a grandmother's or great-grandmother's, since most of the serious seed saving tended to be done by older women. Seeds would be saved from the mutant bean and grown the following summer to see whether they bred true and were tender and tasty. If they were of good eating quality, they became part of that family's seed stock to be kept, even cherished, and shared with kinfolk and others in the community as well.

About twenty-five years ago I had such an experience with a sport myself. A man from Cincinnati stopped by the Lexington Farmers Market to buy some heirloom beans to take back to Cincinnati with him. Buyers of heirloom beans usually want to talk as well, rather than just buying and walking off, and he told me of a bean that he had taken with him from his home in Harlan County,

Kentucky, to his more recent home in Cincinnati. He was so interested in what I was doing with heirloom beans that he made a trip back to Lexington the following Saturday to bring me a handful of his brown greasy beans.

Since it was late in the season but still early enough to plant beans and save them for seed, I planted the beans to develop my own seed stock of his beans. And because they were the last beans I planted that summer and at least three weeks later than any of my other beans, they grew in isolation, which rules out the possibility of their having crossed with other beans. To my surprise, one of the beans was two weeks earlier than the others and had hulls at least two to three times the length of all the others. In addition, it was a white bean while all the others were brown, just like the seed beans I had planted. And while all the others were greasy beans, the new bean had the fuzz of typical cornfield beans on the surface of the hulls.

I carefully gathered all of the beans from that one plant without even shelling them out, put them in an airtight plastic bag, and placed them in a freezer, where I kept them for eighteen years, postponing planting them. I finally decided I had to see what I had, but I chose a bad time, with a rainy period ensuing just after they had been planted. Of sixty-one planted seeds, only nineteen survived, while the others damped off. The nineteen plants produced enough beans for our family to have a good meal and enough seeds to plant a full three-hundred-foot row the following summer. The beans are very tasty and have a very tender texture.

In keeping with naming traditions for family beans, I named the new bean the Robe Mountain Bean, after the mountain behind our house. And for the past several years, I have made the seeds available to others through our website. I have great respect for family beans and see them as important contributions to the genetic diversity of beans. I also feel lucky to have been present when a new bean came into being. It gives new meaning to the phrase "God Given."

Community Beans

Beyond family beans, there are many other beans that have come to predominate in a given community. Because people from several families often shared in the stringing and breaking of beans for canning or making dried beans, they would also swap stories about their own favorite beans. One thing would lead to another, and beans would be swapped to be grown by others in the next growing season.

More than ten years after my mother's death in 1992, my youngest sister, who had moved back home from Charleston, South Carolina, to live in the house in which we were raised, suggested that I ought to look in Mother's freezer, which was kept in our can house along with her jars of vegetables, fruits, and meats. In the freezer, we found thirteen varieties of beans, some of which she had been given by neighbors and cousins.

During the last three years of her life, she had not grown large gardens and had not planted all of the beans that had been given to her. One of the beans, the Lazy Daisy, in particular intrigued me, because it had come from my father's first cousin, who would have been in his mideighties when he gave the variety to her. I already had two varieties of Lazy Wife Greasy beans from Madison County, North Carolina, acquired from the cousin of one of my first cousins on her mother's side. (Extended families can be very helpful and also hard to keep up with.)

The following summer, I grew all thirteen varieties found in Mother's freezer, and all thirteen did well. The Lazy Daisy was a beautiful, medium-sized greasy bean and one of the best-tasting varieties I had ever grown. All the varieties were in a freezer for over twelve years but remained viable. The Lazy Daisy beans might actually have been there for over fifteen years, since they were in the original container, which was still full.

The main point is that bean varieties such as these tend to circulate among people within a community who see one another at bean stringings, family reunions, church suppers, or dinners-on-the-ground. At all such events where meals are served, beans are an important part of the meal, served fresh, frozen, pickled, or as shuck beans. If the beans are good, many people within a given community will grow them.

County and Regional Beans

Seed swapping also extends beyond communities, and many beans tend to be swapped throughout a given county. Such events as court days, rural electric cooperative meetings, revivals, political get-togethers, farm tours, harvest festivals, and other all-county events become good places to swap seeds as well as fish tales. In my home county, farm store operators and hardware stores have also played a role in spreading beans around, especially by selling seed beans brought in by customers in trade for other items, a custom that is still active. For example, J. B. Mullins, a noted bean grower in Breathitt County, Kentucky, trades many varieties of beans to a local hardware store in exchange for many of his supplies for the upcoming summer. The store then sells his beans to its other customers until the supply runs out—always quickly.

Over many years, some beans have become so popular that they have jumped across county lines and have become favorites in much larger geographic areas. The Goose Bean, for example, is known throughout the Southern Appalachians. While the predominant Goose Bean is a deep beige, about six to eight inches long, and typically with a pink tip at maturity, other beans are called Goose Bean in scattered areas. But when most people talk of the Goose Bean, they mean the deep beige one.

Another regional bean, though it does not have quite the reach of the Goose Bean, is the Turkey Craw Bean. This bean is very popular within about a hundred-mile radius of Cumberland Gap (where Kentucky, Virginia, and Tennessee meet). Turkey Craw

Beans are especially popular from there to Kingsport, Tennessee, and throughout Lee County, Virginia, and Harlan and Letcher Counties in Kentucky. Most sources have the variety originating in southeastern Kentucky, but no one knows for sure.

Still another regional bean is the Paterge (Partridge) Head Bean in Albany, Kentucky, and Byrdstown, Tennessee, as well as surrounding areas. It is a large bean in comparison to the size of the hull and is good both as a green bean and for shelly and dried beans. It is a light brown color with darker stripes and grows to be about six inches long.

The Big John Bean is well known in the Knott/Perry/Letcher/Harlan County areas of southeastern Kentucky and is also in high demand among natives of the area who have migrated to Ohio, Michigan, Indiana, and elsewhere in the north. The prices of such beans sometimes climb to $70.00 per bushel on the side of the road on summer Sunday afternoons in July and August when those heading back to their homes in other states stop to buy beans for eating fresh, canning, and drying. At the same time, some of the best strains of the Big John are grown in those same northern states by people who took their seeds with them when they migrated.

Other regional beans include the Logan Giant in southwestern West Virginia and the Mountain Climber in northwestern North Carolina and upper East Tennessee. Of course, there are several greasy beans that have been made very popular by the Western North Carolina Farmers Market in Asheville; many of those bean varieties originated in Madison County, North Carolina, but have become so popular that they are now grown in adjoining counties.

These well-known regional beans are probably among the oldest of the heirloom beans in the Southern Appalachians and have been around for generations, while those originating more recently from crosses or mutants, though no less tender or tasty, have not had time to spread much beyond where they originated, despite the mobility of our modern society. Contrary to the dominant academic opinion that the people of the Appalachians lived in geographic isolation, mountain folk did a lot of traveling and a lot of trading along the way, long before the advent of a cash economy and motorized transportation. It was nothing to take a trip of several hundred miles and be gone two or three months or longer at one time.

Two groups of widely traveled people, politicians and preachers, are given much of the credit for the early spreading of bean varieties, and there are several varieties of preacher beans (although I do not know of any being called "politician beans"). Appalachian scholar, storyteller, and humorist Loyal Jones tells a story that might explain the dissemination of many bean varieties in western North Carolina via politics:

> Zebulon Vance, Civil War governor of North Carolina and later senator, as good a politician as there was, knew the importance of beans, and he used them in his political campaigns: He'd have his wife tie up a few seed beans in a packet, and when he went through the counties campaigning, stopping at houses along the way, he'd say to the woman of the house, "My wife wanted you to have some of her Lazy Wife beans, and she wondered if you could give her a few of your Goose Craw beans." The woman would dutifully tie up a few of her beans, and he would take them to the next house and say, "My wife wanted you to have some of her Goose Craw beans. . ." And so on.

Trade was another mechanism for spreading bean varieties. For example, people going on periodic trips from the mountains to the coast to boil down seawater for salt might have taken beans with them to trade along their routes. I remember as a small child hearing some of the oldest men in the community where I grew up talking about their trips to the ocean to boil down salt. One might then say that a good collector of regional beans would be worth his salt. (An old saying in the mountains holds that someone who is motivated and productive is worth his salt or, if he is nonproductive or lazy, isn't worth his salt.)

Some of the sons and even a few daughters from large Appalachian families became migrant farmworkers during the summers in the early decades of the twentieth century. They traveled on freight trains to Georgia or Florida and followed the harvest from there northward to Pennsylvania and New Jersey and then came back home to the mountains when the harvest was finished. At times, young couples would go work together as migrants and sometimes decide to stay on one of the farms where they did harvesting work. Some, including couples from my home community, made enough money to buy farms of their own in other states. Long before farmers became dependent on migrants from Mexico and other countries, there were sufficient internal migrants to take care of farm labor needs in the United States. My father was one such migrant for many years, spending a lot of time in the summers in New Jersey when it truly was the "Garden State."

Types of Appalachian Heirloom Beans

Cornfield Beans

For some decades, many farm and hardware stores in the mountains have sold bean seeds labeled as "Genuine Cornfield Beans." Such a label is merely a marketing ploy to attract those with little knowledge about beans, or as one might say, to attract buyers who "don't know beans about beans."

It is safe to say that well over 95 percent of Appalachian heirloom beans are genuine cornfield beans. This simply means that they are climbing beans, for which cornstalks historically served as poles for the beans to climb. Some of my most vivid early childhood memories involve going to our cornfields with my mother to pick the beans from the cornstalks when they were at their peak for canning and drying. I thought that the cornfield must be the closest thing imaginable to a jungle. I was fascinated by the many colors of bean hulls and the multicolored beans within them as well.

As hybrid varieties of corn with their shorter and weaker stalks entered the picture, many gardeners started using poles on which to grow their beans, often in teepee style to stabilize them during windy weather. Others would cut tree branches with many twigs to create a bean vine that looked something like a large fan; the bean would continue to send out new vines and be very productive over a long season. Cornfield beans thus became pole beans.

But not all gardeners switched over to poles to support their beans. Many still think that the older open-pollinated corn varieties with their taller and stronger stalks are the best supports for climbing beans. Many gardeners maintain their stock of open-pollinated corn for beans to climb and also for grinding into cornmeal, because in many areas hybrid corn is thought to produce inferior-flavored cornmeal. A popular open-pollinated corn still sold in farm stores is the Hickory Cane (sometimes called Hickory King) or Eight Row variety.

Many growers now grow their climbing beans on trellises supported by strong posts and wires. This is especially true for those growers who grow heirloom beans for farmers' markets, which are springing up throughout the Appalachian region. Such trellises allow for the greatest amount of sunlight on the leaves of the beans and for drip irrigation, which has greatly assisted bean growth during the dry summers of the past few years. Some gardeners also use concrete-reinforcing wire stacked two rolls high and supported by steel posts to create strong trellises. Trellises are becoming increasingly popular with growers, and climbing beans might someday be called "trellis beans" as a more accurate descriptive name.

Runner and Half-Runner Beans

I am often asked this question: "What is the difference between a half-runner and a full runner bean?" My reply is, "About ten feet." The answer is actually more complex, since running beans have runners of many heights. The Peanut Bean, also known as the Pink Half-Runner and sometimes as the Six Week Bean, has runners about three feet long. Some growers contend that the true Peanut Bean has no runners at all, but I have never seen one that completely lacked runners.

Other beans, commonly known as "half-runners," have runners up to at least ten feet. Full runner beans have probably never been accurately measured. I have experimented with bean posts sixteen feet tall on which the bean vines went all the way to the top and back down partway, limited only by the end of the growing season. I have also had running beans, under optimal temperature and moisture conditions, grow more than a foot per day, by actual measurement. As some people have said, "Watch out or they will run over you." But they are not quite like kudzu.

For many years, white half-runners were the dominant bean in many areas of the mountains. This development did not escape notice by commercial seed interests and farm stores, especially those managed by co-ops, which started selling half-runner bean seeds. For many years the seeds were of fairly high quality, and many gardeners stopped saving their own half-runner seeds, assuming that the commercial seed beans would continue to be of high quality.

However, as seed production became more centralized with fewer growers growing seeds for fewer companies, half-runners were contaminated by the tough gene that had been introduced into commercial bush beans so that they would not break during mechanical harvest. As a result, gardeners are having to discard more than half of their beans because the beans are too tough to be edible. I have been told by several farmers' market customers that they stopped buying half-runner beans when they had to throw away more than half of them.

Fortunately, there are still several heirloom half-runner varieties in isolated locations, and many gardeners have sought them out and have started saving their seeds once again. I was given some half-runner seeds by a friend a few years ago. As is true of many heirloom varieties, this bean was a three-in-one bean, having three distinct half-runner variants. I have spent the past ten years stabilizing one of the variants, which I call the NT Half-Runner (NT standing for "non-tough"). I still have the other two variants to go but hope to stabilize them as well.

One of my friends has refused to give up on the commercial half-runner and patiently removes each of the tough beans when they first come up. He says that the tough beans have a slightly different leaf structure and can be pulled up easily soon after they emerge from the soil. I hope he starts saving seeds from the beans he does not pull up; he might bring that particular half-runner back to its original form.

Greasy Beans

Greasy beans are usually thought of as being the best of Appalachian heirloom beans. The fact that they command prices far higher than commercially grown beans attests to their popularity: indeed, they bring up to seven times as much per pound as commercial bush beans, which are typically picked at a very immature stage by machine. Such commercial beans, if allowed to form even very small seeds, are usually too tough to eat.

Greasy beans, as mentioned earlier, are so named because they have slick hulls that look as if they have a thin coat of grease on them, and they are exceptionally tender and tasty. Even when they are fully mature and have turned yellow, they can be strung and broken easily. They are excellent when eaten fresh and are in high demand when made into shuck beans.

Greasy beans ready to be picked for eating

Greasy beans are not a variety of bean but a type. The many varieties of greasy beans come in many colors and many lengths: there are those only two to three inches long and others six to eight inches long. Most of the shorter ones seem to be cut-shorts, while the longer ones have wider spaces between the beans without the beans even touching. All greasy bean hulls are so thin that when held up to the sun, the beans inside are quite visible.

Greasy beans ready to harvest for seed

Greasy beans at the shelly stage

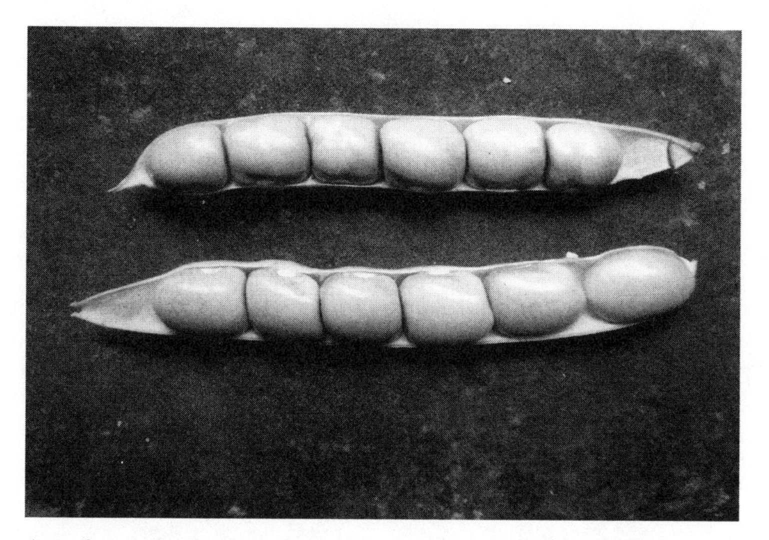

A perfect example of a cut-short bean. *Photo by Michael Best*

Cut-Shorts

When Appalachian heirloom beans are discussed, the term *cut-short* often causes confusion, leading many people to call them "short-cuts" instead. Others think that cut-shorts and greasy beans are one and the same, but the term *cut-short* simply describes what has happened in the hull as the bean grows: the beans grow large in proportion to the hull and tend to square off on the ends.

As cut-shorts dry on the vine prior to being saved for seed, another interesting thing sometimes occurs. After becoming partially dry on the vines, if a sprinkle of rain comes (or even just a heavy dew), the beans can swell up and break the hull open. Because of this, some traditional growers call cut-short beans "bust-out" beans instead. Whenever someone asks me if I know about bust-out beans, I know they are talking about cut-shorts.

Fall Beans or October Beans

The terms *fall beans* and *October beans* are typically used interchangeably. I will use *fall beans* in this discussion.

To ensure a full complement of beans for eating fresh and for preserving for later use, many, if not most, traditional gardeners plant a row or two of fall beans in addition to their other cornfield beans. Fall beans are typically larger in the hulls than other beans are, and the beans are often stringless (most other heirloom cornfield beans are not stringless). Fall beans also lag some two weeks or even longer behind most other beans in maturing: this late maturity is how they came to be known as fall beans.

While the hulls of fall beans are somewhat tougher than other cornfield beans, most are usually tender enough to be eaten with the mature bean seeds just as one would eat green beans. However, if they are to be eaten fresh, most people shell them out at the "shelly" stage (before they have dried) and prepare them as soup beans. Since the advent of refrigeration, many people also freeze them and put them in airtight containers for eating later.

Other fall bean enthusiasts allow their beans to dry on the vine and then shell them out as dry beans to be eaten in much the same way as one would eat pinto beans, cranberry beans, or commercial horticultural beans. Of course, it is best to rehydrate them by soaking them in water for several hours before cooking them. The advantage of eating or freezing them at the shelly stage is that then they do not have to be rehydrated. In addition, most growers believe that they have a better flavor and texture at that stage.

Fall beans come in many colors, from solid white to solid black and all colors of the rainbow. Some are speckled, some are striped, and some contain both speckles and stripes. Some fall beans are named after their interesting color patterns. One example of this is the Baby Face fall bean, with a pattern that looks like the face of a baby; a friend of mine found this variety for me while he was traveling in southeastern Kentucky a few years ago.

Many fall beans are stringless, but a few have strings. Their seeds are typically rounder than those of most other beans. Some have eye colors different from the rest of the bean. Most are climbing beans, but a few are bush beans. Those that are bush beans tend to be stringless, but there are exceptions to this rule as well.

Pink Tip Beans

Pink tip beans have a pink tip on the blossom end when they become nearly mature. As the seeds become fully sized, the pink tip becomes very obvious and a sign that the bean is ready to be picked from the vine and eaten fresh, canned, or dried.

East Tennessee seems to have more pink tip beans than other parts of the Southern Appalachians. For two years running (2004 and 2005) I attended the Farm Expo in Kingsport, which features farm machinery, grafting demonstrations, a host of craft and food exhibits, many types of 4-H and Future Farmers of America (FFA) projects, talks and demonstrations by agricultural experts, and a lot of entertainment. I attended as an heirloom seed collector and seller.

I also did a lot of trading of beans with many of the old-time gardeners who showed up both years. (When they saw my beans, they brought theirs to the Expo the following day.) I had grown up with my mother's white-seeded pink tip beans, but there were many varieties of pink tip beans that I had never seen before, with most of them being brown-seeded instead of white. When I later grew them out, I came to realize that most were much larger than the white-seeded varieties I already knew about.

I also became aware about that time of a variety of greasy beans that has a pink tip. A gardener from my home county sent me some pink tip greasy beans that I found to be quite good. They were about two weeks later than most other greasy beans, and the pink tip appeared just as they reached complete maturity, a day or so before the green hulls began to turn yellow and at a time they needed to be picked unless they were to be kept for seed.

Stringless and Three-String Beans

I rarely have anyone ask about a stringless heirloom bean, since most people have such a preference for string beans. Stringless beans tend to have a tougher hull than string beans do, which means that they have to be picked earlier, before the seed matures. However, many, if not most, fall beans are stringless. And while their hulls are somewhat tougher than those of string beans, many people still break them and eat them as green beans. The beans separate readily from the hulls during the cooking process because the beans are so large in proportion to the hulls.

Some varieties of beans have three strings, one on the outer side and two on the inner side of the bean. (The inside of the bean is the side within the curl that almost all heirloom beans have.) The two strings on the inner side of three-string beans are side by side, each "zipping" its own half of the bean pod, and the two strings peel off easily together.

Wax Beans

Although wax beans are not widely grown in the mountains, many gardeners keep at least one variety. Wax beans are pale, from yellow to almost white. Some people use them in three-bean salads, and some cook them for use in the same way as green beans. They range from very thin hulls to exceptionally thick hulls and tend to have a less "beany" flavor than do green beans or purple-hulled beans (which turn green when cooked).

Butter Beans

Many Appalachian gardeners also keep at least one variety of butter bean, ranging from white to deep purple in color. Many are striped or speckled. They do not cross with green beans or wax beans, but they do cross readily with one another. Their hulls are not eaten, and they are much later than green beans in maturing, sometimes taking 120 days, so they must be planted early to ensure that they reach maturity before frost. I do not know of any heirloom bush butter bean grown in the Southern Appalachians; the heirloom varieties require cornstalks, poles, or trellises to yield effectively. They can be shelled prior to the hulls becoming dry as shelly beans and may be eaten without having to be rehydrated, but most people tend to eat them as dry beans because they store well.

Butter beans often come with stories attached. One Kentucky gardener has butter beans that can be traced back to the end of the Civil War. His great-grandfather was near New Orleans at the end of the war and had to walk back to Kentucky. Whenever he passed a garden where butter beans were being grown, he would collect some of them, and he ended up with an amazing array of colors. In retirement, Joe Richards keeps growing them at his home in Somerset, Kentucky.

Types of Heirloom Beans in Tennessee

John Coykendall, master gardener for the Blackberry Farm Resort, has long experimented with heirloom and heritage vegetable seeds. In the following two sections he discusses the many old-time bean varieties he has collected.

Collecting Beans in Tennessee

JOHN COYKENDALL

During the years that I have been collecting beans in Tennessee, it soon became evident that the vast majority of old varieties to be found were in rural counties in or near the Smoky Mountains region of East Tennessee.

It was in isolated coves and hollows that I found most of the old bean varieties that I collected. Families in these regions had saved their own unique varieties down through the years. Often these old varieties had special characteristics that had developed over time, through selection and isolation.

There were beans for each growing season; some tolerated early spring planting when the soil was still cold, while others were grown during the hot summer months, and these included stick beans, pole beans, and cornfield beans, which were somewhat shade tolerant and suited for growing in cornfields. There were also a number of fall or October beans that thrive during late summer and early fall.

Most people are familiar with the dry shell types of horticultural beans, which have tough hulls at all stages of development. Not many people, however, are aware that there are a number of tender-hull October bean varieties that were once commonly grown.

I have five different October beans in my collection that come from Campbell and Greene Counties here in East Tennessee. With the tender-hull October beans you had a multipurpose bean; they were good to use at all stages of development, including fresh shell beans. In late fall the dried pods were gathered and shelled out to be used as dry beans during the winter.

One of my earliest memories of beans was "leather britches." I remember seeing them hanging on long strings from rafter poles, on the front porches of farmhouses, or from nails on the walls of back porches. I especially remember a neighbor lady in a calico dress with a large apron and wearing a split bonnet. She was sitting on the front porch of her cabin stringing up green beans to be dried for winter use.

Today a few people still string up leather britches, although the necessity of doing so is long past. For some it is a part of a nostalgic tradition that is still being carried on. Perhaps for many it is carrying on what they remember their parents and grandparents doing. For others it may be the unique old-time flavor that awakens memories from long ago. For me leather britches represent a celebration of our culinary traditions, along with history, heritage, and a way of life that is unique to our mountain region.

As seed savers we are not only preserving old varieties; we are also keeping them alive by "using" them, selling them at farmers' markets, and introducing them to chefs who are always on the lookout for something unique for their culinary creations.

My personal favorite method for preparing leather britches is to cook them in a cast iron pot over a slow fire and season them with a piece of smokehouse meat. They are also excellent with potatoes cooked on top of the beans, and the addition of a potato onion also enhances the flavor. In this region, greasy and cut-short beans were commonly used to make leather britches, although a number of other types were also used.

A Few of the Old Tennessee Bean Varieties

Occasionally when collecting seeds I will be given a variety with a name that suggests its origin. Two examples come to mind—the first being Old Time German, which is a strikingly beautiful bean with pods that are light green in the early stages of development and light pinkish violet when fully mature. The elongated seeds are light pink-gray-tan in color, and the flowers are a faded pink-violet hue, making a beautiful addition to either flower or vegetable gardens. The second example is called Old German and is similar to the white half-runner types. During the mid-1800s, a good number of Germans settled in East Tennessee, so they may have brought seeds from these beans with them from Germany. Both examples are pole beans.

Most of the old bean varieties that I have are pole beans, and that seems to be the case down through the history of our region. One of my old mountain friends, Herb Clabo of Sevier

County, Tennessee, who is now one hundred years old, once told me, "If hit's worth havin' it's worth stickin'," and I have found that to be true for my preferences.

Although the greatest diversity of bean varieties is to be found in the mountains of western North Carolina and the eastern sections of Kentucky, in East Tennessee, especially in the mountain regions, a good number of old beans are still to be found that have been grown, preserved, and handed down through the generations.

To a lesser extent there are still some old varieties being grown on the Cumberland Plateau, but in West Tennessee where the country is flat and large-scale mechanized farming has been practiced over a long period of time, old varieties fell out of favor and were replaced with modern varieties. This is not to say that old varieties don't exist in West Tennessee; it is that their numbers are fewer, far between, and difficult to locate.

Below are listed some of the more unusual varieties that I have collected over the years. With the exception of a number of white-seeded beans, the vast majority of beans have beautiful seed coat mottling and are works of art worthy of display.

Milk and Cider—One of my personal favorites is a pole bean called Milk and Cider that came from Claiborne County, Tennessee. The green pods are from five to six inches long with slightly curved pods. The beans' seed coat mottling resembles the Turkey Craw Bean, with the exception that the light gray color appears as though it had been airbrushed onto the seed coat. As is the case with many of the old bean varieties, Milk and Cider remains tender at all stages of growth, right up to full maturity when the pods are well filled out.

Southern Cornfield—The Southern Cornfield bean was collected in Sevier County, Tennessee, and was once commonly grown in cornfields across Southern Appalachia. The pods are semiflat, measuring six to seven inches in length with the pods being slightly curved. The elongated seeds are light tan with dark brown stripes. The Southern Cornfield is a prolific producer, often producing eight to ten beans to a tag.

Mountain City Whitehull—One of the most unusual beans in my collection is the Mountain City Whitehull, which comes from upper East Tennessee. The pods measure from five to six inches in length, and the well-filled-out pods are white with a hint of light yellow. The seeds are white and range from medium to large. This is the first white-hull bean that I have been able to find, although a good number of varieties were once available.

Old Time Golden Stick—Collected in Fentress County, Tennessee, in 1994, the Golden Stick Bean is an excellent all-purpose variety that matures early in the growing season. This bean very much resembles the white half-runner and makes a good substitute for it. At full maturity the seeds are golden tan.

Pink Tip Pole Bean—Pink tip beans have long been popular in upper East Tennessee. This variety is a heavy producer of five-inch-long beans that are an off-yellow to white color fading to a pink hue at their tips, and finally turning to a pink-purple at full maturity. The seeds are dark tan. My seed stock came from Unicoi County, Tennessee.

Red Goose Pole Bean—The Red Goose Bean is a heavy producer of six- to seven-inch pods that are well filled out at maturity. When the beans are nearing full maturity, the pods fade to a pinkish-violet hue. The seeds are dark red and elongated in shape.

Pumpkin Bean-Pole Variety—The Pumpkin Bean produces large pods that are from six to eight inches in length. Two or three of these and you have a plate full! The large elongated light tan seeds are quite large and have a wonderful flavor when the green beans are cooked after the pods have matured.

Butter Beans

People often ask, What is the difference between butter beans and limas? The answer is, They're all limas. Outside of the South, when the term *lima* is used, it generally refers to the green limas and in some cases the large and small white varieties. Here in the Southland it is the speckled limas that are referred to as butter beans, with Florida Speckled and Jackson Wonder being two good examples. Although butter beans are grown to some extent in East Tennessee, they are not nearly as common as they are in the Deep South, where no large garden would be complete without them.

In collecting butter beans, it soon became evident to me that the diversity among them rivaled beans in terms of diverse colors, seed coat mottling, and culinary uses.

My favorite butter bean is the *Red Calico*, which has been preserved by the Thweat family in Tennessee since 1794. If a number of jars filled with different butter beans were on display, your eye would immediately be drawn to the Red Calico.

Some of the beans are solid red, while others are red with black mottling. In times past when several of the red-seed varieties were grown, it was often considered to be a poverty variety. I would rate it far above the "poverty" level, for it has excellent flavor as a green shell bean and also as a dried bean for winter use. It could be said that this all-purpose bean could single-handedly stave off starvation, because it can easily last from one season to the next.

One of the beneficial traits of this bean is that it produces two crops in one season. During the summer months, the vines are heavily laden with more beans than a good picker can keep up with, which leaves an abundance of dry pods to be picked and shelled for seed and as a food source.

Along about mid- to late August, the vines cease blooming and lapse into a period of rest. Then, just as abruptly as they stopped, they spring to life again, producing new growth, blossoms, and another heavy crop that continues on until the first killing frost, which can be well into November.

Snow on the Mountain—Another beautiful example of one of nature's "artistic creations," this butter bean has been grown since at least 1840 and was referred to as the Speckled Sava during the early years of its history. When one looks at the seed coat mottling, it quickly becomes evident where its adopted "nickname" Snow on the Mountain came from: the seeds are deep maroon with white feathering at the top, giving the appearance of snow on a mountain peak. In growth habit, Snow on the Mountain is the same as the Red Calico and can climb to a great height.

Field Peas

Many farmers in East Tennessee raised field peas, especially in times past; they were grown for soil improvement, green manure, and hay as well as for table use. Over many years of collecting field peas, I have found them to be the same as butter beans in that the vast majority that I have collected came from the Deep South, where many old varieties are still to be found. Three old varieties come to mind that I found being grown in our region: the Whippoorwill, the Gray Crowder, and the Clay Pea.

The *Whippoorwill* dates to the 1830s and has light tan seeds with dark brown speckles and some solid brown seeds.

The *Gray Crowder*, sometimes referred to as the Turkey Crowder, was once raised in the cornfields and was usually planted at lay-by time, an old term that meant you had plowed your corn, cotton, or other crop for the last time and it would not be worked again until harvest time.

When planting peas in the corn, the "middles," or between the corn rows, would be plowed out and peas sown in the furrows for a fall crop. The Gray Crowder is a strong climbing pea and produces large speckled gray peas that are "crowded" together in their pods, producing "crowder peas."

The third example, and once the most commonly grown pea in this part of Tennessee, is the *Clay Pea*, and its well-known history predates the Civil War era. Today Clay Peas are used as a cover crop and for seeding deer plots.

Naming Traditions

Family and community beans are usually named after the person the bean is most associated with. Many have the name of a woman, with the terms *Granny, Grandmaw,* and *Aunt* often attached to bean varieties, while some communities use women's given names or full names. Varieties named after women include Ora's Speckled Bean, Aunt Bet's Bean, Aunt Beth's Bean, Bertie Best Greasy, Margaret Best Greasy, Clova Collins Fall Bean, Grandma Barnett Bean, Vertie's Bean, Anna Robe-Terry Bean, Evelyn Wheeler Pink Tip, Nannie Coulton Bean, Alice Whitis Bean, and Cora Rainey Bean.

I know of only three beans that go by a man's name: the Big John Bean from the Letcher/Harlan County area of Kentucky; the Frank Barnett Cut-Short from Georgetown, Kentucky; and the Grady Bailly Greasy from Polk County, North Carolina. My collection of heirloom beans contains far more named after women than those named after men. This is probably because women were historically associated with seed saving more than were men, but that has been changing during the past fifty years as many men have taken up seed saving, usually brought into the process by their mothers, grandmothers, aunts, and great-aunts.

Regional beans tend not to have the names of individuals as often as do family and community beans. The Goose Bean and Turkey Craw Bean, as noted earlier, have similar origins and are well known in three or more states each. The Mountain Climber Bean in upper East Tennessee and northwestern North Carolina also crosses state lines, as does the Logan Giant, which is well known in parts of eastern Kentucky and West Virginia.

Other beans, such as the Baby Face mentioned earlier, are named after the way they look. The Paterge (Partridge) Head Bean of south-central Kentucky and north-central Tennessee resembles the head of a partridge. The Wren's Egg Bean looks like a wren's egg. The Pink Half-Runner has a pink hull with a deep pink

Jimmie Tallent Sloan with canned Paterge Head Beans

bean and has a short runner. The Case Knife Bean resembles a Case pocketknife. Put two Ram's Horn Beans back-to-back with the crooks on top and you almost have the face of a ram, complete with horns. The Tobacco Worm Bean when mature looks like a fat tobacco hornworm.

Some beans are named after the times they are planted or mature. Six-week beans, for example, mature early and can thus be planted later than other beans. Fall or October beans are slow to mature and need a longer growing season.

Other beans are given names to indicate a particular planting time. Twelve years after my mother died, I found a Tobacco Bed Bean among the beans stored in her freezer. For many decades, farmers who grew tobacco would plant a late crop of beans in their tobacco beds after the tobacco plants had been pulled and transplanted in other fields. Because the tobacco beds had already been burned to kill disease organisms and weed seeds and fertilized to grow the tobacco plants, the beds were ready to plant a crop of late beans or greens without further preparation. It was a good use of fertile soil that was free of disease organisms and weeds and gave the family even more beans for eating fresh, canning, and making leather britches.

I did not remember that Mother's Tobacco Bed Beans were climbing beans, because we never staked them in the tobacco beds and the tobacco harvest came too late to plant corn for the beans to climb on. These climbing beans twined around one another but still produced a lot of beans, although not as much as they would have if they had been given the support of cornstalks or poles.

It is interesting that we have both Tobacco Worm Beans and Tobacco Bed Beans, both excellent beans and both associated with tobacco, a crop that has virtually disappeared in the mountains with the loss of the U.S. government poundage and price support program. Tobacco is no longer a small farm enterprise; it is grown mostly on large farms with most of the labor done by a migrant workforce from outside the country. The bean names remain, however.

Still other beans were given what appear to be whimsical names. There are several varieties of Lazy Wife Greasy beans and at least one Fat Man bean. Both Lazy Wife beans and Fat Man beans are quite large and thus require less effort to pick and prepare for later use: even a "lazy" person would have little trouble growing and picking them.

Some years ago I received an e-mail from a lady in Florida asking whether I knew how she could find some Fat Man beans. Her ninety-two-year-old father, who had been raised in West Virginia, was living with her family in Florida. He had gradually lost his sight and was quite depressed. She thought that tending to some of his favorite beans might cheer him up and give him further reason to live. I had very few of the Fat Man seeds at that time, but I sent her some to plant for her father. Later that summer I received a photograph from her showing her father in his bean row accompanied by his grandchild. He was twining his beans up strings and feeling the developing pods. The beans looked very

Lazy Wife Greasy beans growing to the sky and back again, towering over Sarah Best, the author's granddaughter

Bean medley

healthy, as did he. He also looked happy, and she said that the beans were the best thing that had ever happened to him.

Seed Saving

The most important part of growing heirloom beans is saving seeds for the following year or, if properly stored, for many years.

Some degree of isolation is crucial to keep beans from crossing. Although beans are self-pollinating prior to the bloom opening and, theoretically, do not cross, other factors do come into play, such as cross-pollination through insects and other pollinators. Bumblebees, for example, can aggressively cause beans to cross-pollinate: they rip open the blooms while pollination is taking place and can carry pollen from other beans with them.

Beans can be isolated either geographically or by time, using a range of planting dates. Those with small gardens should place

Completing the drying process before bean seeds are shelled out

their bean varieties as far apart as space will allow and plant some weeks apart if possible. Another way to isolate them is to put light row covers such as Reemay over the bean vines during the blooming period and take them off after the young beans have formed. I have found that some bean varieties rarely cross while others, such as the Goose Bean, must be grown in isolation as much as possible.

Once the beans have matured on the vine and the bean pods have started to turn yellow, the mature beans can be shelled out and spread out to dry. Once dry, they need to be put in airtight containers and refrigerated or frozen. Before freezing them, place the container of room-temperature bean seeds in sunlight for a few minutes during the morning to see whether any moisture appears within the container. If it does, the seeds need to be spread out again and allowed to dry further.

A better way to harvest the seed beans is to let the bean hulls dry on the vine until they are a light brown color and somewhat brittle to the touch. The bean pods can then be gathered and placed on sheets or screens to allow most of the remaining moisture to leave them. If rain is predicted, the bean pods should be gathered prior to the rainfall. Otherwise, if it rains off and on for a period of days, the pods may turn dark and become infected with fungi, which might leave the seeds discolored once the pods dry out again.

Also, if beans are left too long on the vine during rainy weather, the seeds may sprout within the pods. This is especially true of bean pods that are touching the ground. About the only thing left to do if this happens, and it is not too late in the season, is to plant the sprouted beans for a late crop. They will come up quickly. This is commonly done with early-maturing beans and can result in two crops in the same season.

The seeds must be shelled out as soon as possible after the hulls are dry, or bean weevils might start to hatch. Instead of keeping the seeds in a refrigerator or freezer, you can place mothballs or hot peppers in containers with the seeds to control the bean weevils, but keeping them cold is becoming the standard way of preserving viability.

I, for one, plant my bean seeds as soon as I take them out of the freezer. Moisture tends to adhere to the cold beans, which seems to make them sprout faster when they are placed in the ground. I realize that this is not standard practice, but it works well for me.

Preparation as Food

Green beans with corn bread, onions, and butter have made a complete meal for many mountain people over the past 250 years. When eaten fresh, green beans are typically cooked with a cured

Greasy beans almost ready to harvest for seeds

piece of pork in a pot on top of a wood, electric, or gas stove. They may be served twice in the same day and left on the stove to serve the following day, and often for two or more days longer, becoming better each time they are reheated. Potatoes may be added to the pot as well, depending on the tradition of the family. Even after buying an electric stove, many old-time cooks have kept a woodstove in their houses to cook their beans and bake biscuits and corn bread.

Shuck beans are rehydrated prior to being cooked, usually over several hours. An easy way to rehydrate beans is to put them in water overnight and pour off the water around breakfast time. Place them in more water and pour off the second water about noon. Place them in water for the third time during the noon hour and pour off the water about 1 p.m. The beans are then ready to cook as one would cook green beans.

Shelly beans, which do not need to be rehydrated (because they are picked before they dry out), can be cooked as one would cook green beans with cured pork, salted to taste, and with potatoes or an onion if desired, again depending on family tradition. Dry beans need to be rehydrated just as shuck beans do and then cooked in much the same fashion as shelly beans, although the cooking may take longer.

Sharing Bean Seeds

Seed Swapping

Many old-time gardeners share their bean varieties with any and all comers and engage in seed swapping as well. However, this is not always the case: some gardeners try to hold on to their unique varieties as much as possible, often with other people sneaking in at night and stealing them.

Heirloom seed exchange at Best Family Farm: (*left to right*) Bill Best, Rodger Winn, Maria Stenger, Ira Wallace, Susan Kenyon, Brook Elliott, Tony West, Ken Bezilla, Jack Woodworth, Ed Meece, Frank Barnett, Darrel Jones. *Photo by Mary West*

Whereas much of the seed swapping of times past happened at family reunions, holiday celebrations, court days, and other times of coming together socially, many agencies are now sponsoring seed-swapping events. We hosted several such events at our farm on the first Saturday in October for the now-defunct Appalachian Heirloom Seed Conservancy. At the suggestion of several of those who attended AHSC events, the Best Family Farm continued the tradition and still hosts a one-day seed swap each year on the first Saturday in October. We now regularly draw more than a hundred serious seed savers in attendance from eight to ten states, all with a minimum of advertising.

Other Means of Making Seeds Available

In addition to politicians, preachers, and peddlers, country stores were also important in moving seeds from community to community. These stores were sometimes on wheels; before cars and trucks were widely affordable, some store owners moved their wares about in trucks similar to school buses. Their weekly visits to rural homes were eagerly anticipated events.

One farm store owner was selling heirloom seeds more than sixty years ago, and I recently found out that one of the beans given to me more than twenty years ago might have originally been part of such a trade at Shorty Ketner's Farm Store in Waynesville, North Carolina, where my father bought much of his hog feed during the winter. This was before some of the cooperatives that now dominate farm stores came into being.

A few years ago I noticed that the bean seeds being sold by Southern States Cooperative stores, for example, were being grown in Idaho. Seeds sold by large cooperatives such as this are grown in large irrigated fields and mechanically harvested. Because of the large scale at which they are grown and marketed, these seeds can be sold more cheaply, but quality is hard to maintain. This has been especially true with the white half-runner, once the favorite bean in the Appalachian region but now so tough that people interested in canning them tell me that they now have to throw away more than half of them.

Very few farm stores now sell heirloom beans, largely because they have no readily available sources of supply, and even if they could obtain the heirloom varieties, they would have to charge two to ten times as much for them because heirloom beans, being unsuited for machine harvest, have to be picked by hand. I have been approached by one large and several smaller seed companies to grow heirloom beans for them, but because of the hand labor involved, it is all I can do to grow enough for our own sales at our home and on the Internet.

J. B. and Josephine Mullins

Today, finding commercial sources for heirloom beans is difficult, and when they can be found, they seem to be more likely to be found in hardware stores than in farm stores. This is because a few growers of heirloom beans trade their seeds to hardware stores for supplies for their small farms and gardens.

The May 1, 2002, issue of the *Lexington Herald-Leader* carried a story by Dick Burdette about J. B. Mullins, a noted bean grower in Breathitt County, Kentucky. Mullins grows such heirloom varieties as the Goose, Greasy Grit, Daddy, Dogeye, 40 Bushel, and Creaseback beans. He sells them to the True Value hardware store in Jackson, the county seat. The store maintains a list of several hundred bean customers who share the forty or so pounds that Mullins brings to the store. Mullins trades the beans for onion sets, fertilizer, tools, and other vegetable seeds or other supplies for his garden.

I visited Mr. Mullins during the fall of 2009 and found him unable to trade his beans to the hardware store for the first time

J. B. Mullins

in many years. The weather was so bad during the 2009 growing season, with heavy rains and one bad flood, that he had barely enough seeds to keep himself going. He is now in his mid-eighties and still planning to keep growing his heirloom beans, although he threatens that each summer will be his last.

Some farm stores, however, do still occasionally have heirloom beans for sale. About thirty years ago a similar trading arrangement was made with the Southern States Farm Store in Berea, Kentucky, by a gardener in Jackson County, Kentucky. She came into the store with a bucket of greasy cut-shorts and left with many of her supplies for the summer. Her seeds were usually gone within a day, and I got some only because I happened to be in the store on the day she brought in the seeds. Unfortunately, that was before I started keeping accurate records of growers, and the beans entered my collection as the Jackson County Greasy Cut-Short. I still hope to find out the name of the lady and get her story. Her beans were the only heirloom beans ever sold there that I know of, and they disappeared quickly each spring. The store has been closed for several years.

Beans and Appalachian Culture

It is hard to discuss Appalachian heirloom beans without exploring the cultural context that has made this section of the United States such a fertile ground for maintaining heirloom beans for generations. Much of that cultural context is exemplified by my home community, the Upper Crabtree community in Haywood County, North Carolina.

Most of my family on both sides are descended from a Revolutionary War soldier from Scotland and his French Huguenot wife, Joseph and Sarah Vaughn McCracken, who were among the first to settle in the community in the late eighteenth century. They had twelve children, most of whom married and had large families of their own. Most people still living in the community are descended from one or more of those twelve children. Their descendants also spread far and wide, with many ending up in Georgia, Texas, Idaho, Oregon, and Washington.

Upper Crabtree is a community of small farms where most of the people in a thirty-mile radius work for small industries, businesses, and schools. Most still maintain their farms, and most still garden. Swapping seeds, especially bean seeds, remains very much a part of community living in Upper Crabtree. It is a custom that refuses to die out and in many respects, as the food situation in the United States becomes more problematic, is growing stronger.

While in high school and college, I had a summer job (in addition to helping on our family farm) that involved measuring tobacco acreage. The original program was based on acreage, with each farm being allotted a certain fraction of an acre or up to many acres, based on what the farm was producing when the program went into effect in 1938. Under that program, when land was sold, the acreage quota stayed with the farm rather than going with the grower. (This was before the advent of the poundage program, by which the federal government allocated growers a certain number of pounds to grow on their farms beginning in 1971.) For many summers I had the job of measuring all the tobacco patches in my home community. After I completed that job in early July, I was usually also given the job of measuring the acreage allotments in isolated parts of the county that did not merit a full-time "reporter," as we were called.

On one memorable trip, my younger brother (who was helping me) and I forded the Pigeon River and walked two and a half miles to near the top of a mountain where a couple grew a small allotment of tobacco along with their garden crops and a field of corn. The woman, Letha Hicks, was an excellent gardener and grew many of the same vegetables as my mother did, including a lot of cornfield beans. She was about the same age as my mother.

I later learned that the corn crop was not for feeding animals but was being transformed into liquid form by her husband. I noticed that he had been drinking some whiskey, but at the time I did not realize that he was making his own and also marketing it to many regular customers.

About fifty years later, in 2007, I learned from a friend in Upper Crabtree, Yolanda Ferguson, that Mrs. Hicks had died in 1996 at the age of eighty-three and that I might be able to obtain some of her prized bean seeds. Yolanda also told me another interesting story about Mrs. Hicks's beans: she was so secretive about them that she had never shared them with anyone, although several people had asked to be given some to grow for themselves. One man, Whitey Swanger, a musician from the nearby community of Fines Creek, desired them enough that he had gone into her patch at night to remove enough to give them a try. After growing the beans for a year or two, using a toothpick to open the blooms prior to pollination taking place, he then crossed his favorite bean with Mrs. Hicks's bean to create a greasy cut-short bean.

Bean Stories

Bean stories abound in the Southern Appalachians. One of the best known, *Jack and the Wonder Beans*, is by noted author James Still, who wrote this Appalachian version of "Jack and the Beanstalk" in accurate dialect. Many mountain grandparents had their own bean stories to tell to their grandchildren, especially stories of the origins of beans such as the Goose Bean and the Turkey Craw Bean.

I have received information from many people by way of letters, phone calls, the Internet, and places I have visited about older people who have stories to tell about their beans. The following are just a few examples of what I learned from these old-time bean grower/seed collectors.

Cliffie Strong (Owsley County, Kentucky)

Cliffie Strong, a ninety-two-year-old widow when I visited her in November 2007, has a remarkable collection of beans and other garden seeds. Her gardening lasted from 1934 through 2006, when her health began to keep her from working in her gardens. She has recovered enough to maintain a keen interest in her garden and supervise her daughter, grandson, and great-grandson as they make up the only four-generation family I know of who are planting the same heirloom seeds. Their gardens are at three different places in two counties.

In addition to growing gardens for her family, she also "hired out" to do farmwork for other people when time permitted, working for fifty cents per day. She simultaneously did massive amounts of farmwork on the farm she and her husband owned, and she canned and dried most of the food needed by their family. This included stringing, breaking, and drying enough beans to make five bushels of shuck beans per season. Prior to their being dried, this would probably amount to twenty or more bushels of green beans.

Bean seeds that she still keeps in her freezer include Little White Bunch, Big White Bunch, Big Brown Beans, Brown Goose, Gray Goose, White Goose, Striped Goose, White Case Knife, Tobacco

Cliffie Strong, age ninety-seven, matriarch of four generations of seed savers

Worm, Pink Tip, Crane Toe, Golden Hull, Six-Week Cream Colored, Fat Bean (a fall bean), Nannie Coulton Bean, and Dry Weather Bean. The Nannie Coulton Bean is a mutant bean found by her friend Nannie Coulton in her garden.

Cliffie was so satisfied with her own beans that she never started growing half-runners when they gained considerable popularity some fifty years ago.

In addition to her many varieties of beans, Mrs. Strong grew Black-Eyed Peas, Whippoorwill Pea, and Sugar Crowder Pea. She also grew "Old Flat" Mustard, white sweet potatoes, and Irish potatoes.

Harold Wallace (Lexington, Kentucky)

Other families had beans that were of such special significance that they were saved for generations. One such bean is the Tennessee Cornfield Bean, which was given to me by Harold and Bernease Wallace. This bean, a brown bean grown by the Wallace family in Putnam County, Tennessee, for generations, is very tender and best when it is firmly full, when the beans are completely mature prior to the hulls beginning to turn yellow.

Harold Wallace with his family's Tennessee Cornfield Bean

Beans were just a part, albeit a large one, of the Wallace family's farming operation. On one hundred acres they also raised cattle, hogs, chickens, and guineas. Other garden crops included tomatoes, corn, potatoes, carrots, beets, lettuce, turnips, mustard, cabbages, onions, and squashes. They also had peach trees. In addition to canning and drying, they kept many fruits and vegetables in a root cellar.

I came to know the Wallaces at the Lexington Farmers Market some twenty years ago and have grown the Tennessee Cornfield especially for them since that time. Other customers are quite fond of it as well. When Harold first gave the bean to me, he remarked that it was an important source of food during the Great Depression, a common story from older gardeners who lived through that time. Harold retired from the Soil Conservation Service after thirty years, having served fourteen counties.

Ella Mae Barnett (Floyd County, Kentucky)

The following story about Ella Mae Barnett was given to me by her grandson, Frank Barnett of Georgetown, Kentucky, who has been growing heirloom beans for some years and is building up quite a collection of his own:

My grandmother, Ella Mae Barnett, was born in 1896 and always raised huge gardens until 1988; she passed away in 1990. She lived her entire life on Bucks Branch in Floyd County, Kentucky, with the exception of two days her son took her to see his farm in Ohio. (The visit was supposed to be one week but she drove him crazy wanting to get back home.)

I remember her having three garden spots. The two at the house were on either side of the front of the house down to maybe two hundred feet to the creek. There was a walk[ing] bridge from the road across the creek with a path to the front porch of the house. On both sides of the walking path were garden sections, which were fenced in order to keep her dogs and chickens out. However, the dogs and chickens were free to roam anywhere else. Of course, there was no grass to cut in the yard around the house; it was just dirt since the chickens kept everything scratched out.

She had a third garden maybe two thousand feet up the holler on the hillside where she raised nothing but beans. There was no need to worry about deer; if they were ever spotted they would soon be fresh meat. (In fact, I never saw a deer while growing up in eastern Kentucky.) There was a crude barb[-wire] fence around the garden, just enough to keep the milk cow out.

The cornfield beans which she gave my dad which he gave to me after she passed away were maybe twenty beans which she had tied up in a rag. She had gotten the beans years earlier from a distant relative in Magoffin County at the head of the Licking River area.

The first year I raised them I had planted them in Hickory King Corn and they broke the stalks to the ground. I saved nearly a quart glass jar of seed. I had noticed that there were a few tough pods which contained all black beans which I threw away. After I started collecting heirloom beans I wished I had saved those seeds and planted them separately in succeeding years. Perhaps I would have gotten a totally different bean.

Kate Sanford (Haywood County, North Carolina)

The following story is not just about beans; it was written by my first cousin, Leon Sanford, about our grandmother, Kate Sanford. Leon, eight years older than I am and the oldest of her grandchildren, lived just several hundred feet from her house and gardens and saw her on a daily basis. He currently lives in Idaho, having retired there many years ago after a career in the military.

As I remember Gram Maw Sanford, Grangy as she was called by all her grandkids, she was a very proud and strong woman. She took pride in everything she undertook. I know she went to baker's school when she was a young girl and learned well. She had the reputation of being the best cook and baker anywhere around. Her pies, cakes, and cookies were out of this world. Neighbors had a habit of passing by her home at lunch time and were always welcome to share her food.

My grandfather passed away at a young age, leaving Grangy alone. I remember spending the nights with her when I was a young lad and being woken up about 5 a.m. each morning. She always kept lots of chickens, and they roosted in some large spruce trees in her yard. I remember the roosters crowing each morning and waking me up. I always accused Gram Maw of waking up the roosters so they could wake up everyone in the country.

Grangy was also a gardener, and she took pride in having the best garden around. She spent many hours in her garden, and it was always weed free with lots of different veggies growing. I remember spending lots of hours in her garden, helping her plant and weed it. On Sunday her children and grandchildren would visit with her, and before they departed for home, a trip to the garden was necessary. She would load their cars with fresh veggies before allowing them to depart for home. She could also always find some excess in her garden to give people in need.

Gram Maw was one of a kind, very wise, and admired by everyone. I admired her very much and also learned a lot from her. She had lots of grandkids and it was always a pleasure for them to visit with her. Sunday was always her favorite day, and after church her children and grandchildren would gather and spend the afternoon with her.

Much of her time was spent doing home canning. And I remember spending lots of time helping her, especially making pepper and onion relish. I remember very well grinding the pepper and onions for her. Her relish was so good that I still use her recipe today.

RELISH RECIPE:

Grind five each red, green, and yellow peppers with about 10 medium onions

Add 2 cups vinegar and 3 cups brown sugar

Cook for about 45 minutes and can in pint jars

(The recipe is as Leon remembers from Grangy's recipe, and he still uses it.)

My most significant memories of Grandma Sanford likewise have to do with her gardens. I remember as a very young child helping harvest vegetables on Sunday afternoons during our weekly visits. I also remember her garden conversations with her only daughter, my mother. More important, I remember the seed-saving habits she was passing on to my mother, which she later passed on to me. I also remember how beautiful and orderly everything seemed to be in her garden.

Hill Family Bean (Haywood County, North Carolina)

Another bean story from my home community was published by the *Mountaineer* newspaper in Waynesville, North Carolina, on October 20, 2000. The article was written by Kathy N. Ross (and is reprinted by permission).

Haywood County's past can be found in more than old records and photo albums. Try the bean patch. It's in the gardens, fields and orchards here that living history can be found in literal fashion, in strains of apples, flowers and vegetables handed down from generations past. Family strains of green beans are among the most common of our horticultural heritage, perhaps because of their hardiness. Take the case of the Hill family beans.

Ben Best's mother was a Hill, one of 12 children who grew up with beans grown by their mother. Though Ben's mother, Cecile Hill, died in 1991, he and his wife, Clarine, found part of that living legacy when cleaning out her canning house. On the shelf rested a half-gallon glass jar of those beans, dried out for seed.

The Bests planted the beans and many of them came up. They saved and dried that crop, and planted them again. This is the third year they've grown the family bean since its rediscovery.

"We were figuring up when we found those beans, and it had been about 20 years since she (Ben's mother) had grown any," Clarine Best said. "They'd sat on the shelf in her canning house all that time." Cecile Hill told her children she had gotten the bean from her mother, which would mean more than 100 years it had been in the family.

The bean is a colorful one, in and out of the husk. As it matures on the vine to the point where the beans are better shelled —"shelly beans," people here call them—the outer husk turns hues of rose and yellow, similar to the shades on a fall maple tree.

Inside, the beans themselves are also colorful and varied. Some are the traditional white; others are striped; some are deep purple while others are black and brown. When they're canned, however, the color fades.

Ben's maternal grandparents, Charles and Lily Hill, had eight girls before giving birth to their first son. Their last four children were all boys. Ben's mother told him how the two oldest girls would plow and cultivate the field by horse, and the youngest girls would follow with a bucket of the beans, digging out holes with their toes to drop them in.

At the end of the season, "they would spread them out on the floor to dry, then thresh them," Ben said. They would put the beans in a 15-pound sack, suspend the sack, then hit it to shatter the dried hulls and free the bean seed.

Venice Davis of Iron Duff, a first cousin to Ben, also grows the colorful cornfield bean. She likes it for its color, its flavor and the fact that it grows hardy and plentiful, she said. One member of the family didn't like the dark purple beans, Davis said. "They would pick the purple ones out (of the seed beans), and when they came up they still had purple ones mixed in; they'd still come on," she said. Davis remembers her mother going up atop the mountain above Upper Crabtree in late fall and coming off with a bushel's worth of beans on her back. "They would tell her the beans were gone, but she could always find a bushel of beans," she said.

The beans have a good taste, said Clarine Best, a little heartier than the popular greasy cut-shorts.

Traveling Beans

Having a website on the Internet has proven to be a godsend for obtaining information about beans, tomatoes, and other heirloom fruits and vegetables. While getting information from me, if and when I have it, many writers have shared valuable information as well. Their stories about beans are particularly informative about the ways beans have traveled from place to place and state to state and about the roles beans have played in the lives of people. A lot of people in their later years seem to develop a yearning for the foods they knew as children. One might also say that they redevelop a hunger for real food, in contrast to the synthetic processed foods so prevalent today.

This is a letter from Sandy Sherwin dated February 8, 2008:

> I wonder if you can help me find a specific heirloom bean seed that I remember my grandfather growing when

> I was a child. He called them "cornfield beans" and saved the seeds from year to year. They were a pole bean with pods approximately 6–7 inches long. I don't remember what color the beans inside were when picked but after snapping and cooking them, the outsides of the pods remained green and the beans inside were a solid brown color on the outside, whitish inside. If I remember correctly, when dried for seed, the beans resembled a small version of a pinto bean. They had a better flavor than any green beans with white seeds that I've ever tasted. Oh, he brought the seed with him from Walhalla, South Carolina, to Oregon in the 1920s. I would love to find some seeds to once again taste these delicious green beans. Thanks for any help you can give me.

I answered Sandy's letter the following day, February 9, 2008, at 4:52 p.m.

> Hello Sandy,
>
> I don't know if I can help you find a specific heirloom bean seed but I have several cornfield beans that are similar to the one you describe. The one most similar to the one you describe is the Edwards Bean which is a brown striped bean which cooks up brown. Another is the Tennessee Cornfield Bean which is a deep beige bean which also cooks up brown. Another might be the Brown Tobacco Worm Bean. The Turkey Craw Bean is brown speckled and an excellent tasting bean while the Goose Bean is brown and seven to nine inches long. All of these beans are favorites of my farmers' market customers.
>
> Do you live in Oregon or have you moved back south? If you have relatives in Walhalla you might ask some of them to see if they might find the bean. Dr. Bradshaw at Clemson University has worked with heirloom beans for some years and he might also know of a bean similar to the ones you describe.

I hope this information is helpful to you. Also, do you mind if I quote your letter in a book I'm writing on Appalachian heirloom fruits and vegetables for the Ohio University Press? Your letter is like those I often receive when family members have lost seeds and wish to go back to the foods they grew up with. This is understandable since modern plant breeding has given us tough beans which can be mechanically harvested but which have no nutrition. Many people from the South who migrated to the Pacific Northwest took their beans with them because they knew something good when they ate it. My aunt and uncle migrated to Kelso, Washington, about 1918 to work in the timber industry and came back to North Carolina only once after that in the 1940s.

Thanks for your interest.
Bill Best

Sandy's response came back the same day five hours later.

Thank you for your quick response. I think the Edwards Bean sounds most like what I'm thinking of from appearance. Before I try ordering any seeds from you, I am going to pursue trying to contact some family to see if I can find the same bean. I no longer live in Oregon and very few of the family there are still living. I do have contact with one cousin whose father also probably raised these beans (my grandfather's brother). I've written to her in Oregon to see if she might give me the name of some family contacts in Walhalla or nearby. I live in Colorado and have only driven through Walhalla once (as curiosity to see where my Mom was born) when we were traveling from Atlanta to Greenville SC where my sons attended college. I would be happy for you to include my note in your book if you find it useful.

Even though I don't garden much now (Colorado Springs isn't quite the right climate compared to Oregon's Willamette Valley or South Carolina) I would like to pass the heritage on to my daughter who has a great garden in Wisconsin.

I may also try to contact Dr. Bradshaw at Clemson if the family route doesn't work out since I know Clemson is just down the road from Walhalla.

Thanks again . . .
Sandy Sherwin

Archaeologists spend a lot of time trying to determine the historic migration patterns of people and their food plants. In a nutshell, Sandy Sherwin has described the most recent travel patterns of one bean variety. The bean started out in Walhalla, South Carolina, and then moved to Oregon. If it can be located back in South Carolina, it will now be grown in Wisconsin by the daughter of a woman born in Washington who now makes Colorado her home. Sandy's attempt to find the bean is to share her heritage with her daughter, a heritage that, with luck, was only temporarily disconnected.

Beans seeds travel easily with people. The Cherokees took at least some of their bean varieties with them when forced by the federal government to leave the Southern Appalachians for the Indian Territory. Southern Appalachian mountaineers took beans with them when, in the early part of the twentieth century, they migrated to Texas, Idaho, Oregon, and Washington in search of employment and land. Many ended up working in the timber industry in Oregon and Washington as they had previously worked in the timber industry in northern Georgia, western North Carolina, or East Tennessee.

There they grew beans from western North Carolina, and the adapted beans came to be known as Tarheel Beans, after the state of North Carolina's nickname, the Tarheel State. Although I have not been able to authoritatively determine which beans are now known as Tarheel Beans, many who know of them believe they are greasy beans from the Haywood/Jackson County of western

North Carolina, which is the most concentrated geographic area for growing greasy beans.

Southern Appalachian Heirloom Bean Terminology

"Old-timers" speak of people who "don't know beans" about a particular subject, meaning that such people know very little about what they are talking about. Because gardening is not a part of the lifestyle of many people today, we increasingly also have a problem of people "not knowing beans about beans." The following definitions are put forth to inform people and to assist them in buying bean seeds for their gardens or purchasing fresh beans from growers or other sellers of beans. (It is a sign of the times that most of this information was common knowledge only fifty years ago.)

Bush or Bunch Beans—There are very few heirloom bunch beans in the Southern Appalachians, but there are some quite good ones. Several are fall beans and have strings, while a few are stringless. As a general rule, bush beans have tougher hulls than cornfield beans do, which lessens their desirability. The plants also produce far fewer beans, making them less attractive to growers with limited space. Depending on variety, they can be broken and cooked when full or while pods are still tender if they are varieties that become tough when full. Or, as with cornfield beans, they can be eaten as shelly or dry beans as well. Most people feel that stringless beans have less flavor than ones that require stringing.

Butter Beans—Butter beans, very common in the Deep South, are grown extensively in the Southern Appalachians as well. They are usually somewhat smaller than their cousins, the lima beans, but are very colorful. They must be shelled out for eating, either as shelly beans or later as dry beans. Many families grow at least one variety of butter beans.

Cornfield Beans—This term can be applied to any climbing bean. Cornstalks traditionally served as the poles that beans used for climbing.

Creaseback Beans—These are a type of heirloom bean that has a crease in the outer portion of the bean hull. They are sometimes called creasy beans (not to be confused with greasy beans).

Cut-Short Beans—These are a type of bean in which the seeds outgrow the hulls, locking the developing seeds against one another. This makes them appear square, rectangular, triangular, or even trapezoidal in form. Cut-shorts are in high demand by traditional growers because of their high protein content. They are sometimes called bust-out beans because the dried hulls will often split apart vigorously after the bean pods have dried out and then become wet again by rain or even a heavy dew. This is nature's way of scattering seeds for the upcoming season.

Dry Beans—Any bean can be a dry bean, since the term refers to the dry seeds of beans. Beans can be allowed to dry while in the hull or can be shelled out as shelly beans and then allowed to dry while spread out on a flat surface. If the weather threatens, many gardeners will pick their beans while still in the shelly stage rather than take a chance on the hulls becoming discolored, which might also discolor the seeds. Dry beans are typically rehydrated prior to cooking by being soaked overnight or longer and sometimes having the water poured off several times before being cooked.

Fall or October Beans—These beans typically have large seeds and sometimes stringless hulls. They are often tougher than other heirloom beans but can be eaten as green beans, as shelly beans, or as dry beans; many families always plant at least one fall bean. There are many varieties of bush fall beans.

Full Beans—This is a term used to describe a bean in which the seed is fully mature within the hull and the bean is ready to harvest. Heirloom beans are traditionally harvested at the full

Rose Cornfield beans from the Rose family of Panola, Kentucky, grown by Earl Bowling. *Photo by Earl Bowling*

stage whether they are to be used fresh, canned, or pickled or for making leather britches.

Greasy Beans—This name is given to many heirloom bean varieties for which the pods are slick and lacking the abundant fuzz of other beans: the slickness makes them appear to be greasy. Greasy beans are widely thought to be the highest-quality beans and are by far the highest priced. Most greasy bean varieties are found in western North Carolina and eastern Kentucky, but they are spreading rapidly to other areas through farmers' markets and heirloom seed outlets. Greasy cut-shorts are in very high demand.

Half-Runner Beans—This is a term given to many varieties of beans for which the runner is roughly from three to ten feet long. It might be more accurate to say that there are quarter-runner beans, half-runner beans, and full runner beans, with full runner beans climbing to twenty feet or more. Half-runners are very popular in the southern mountains, and this led to commercial seed companies starting to produce and sell seeds. This further led to the tough gene being implanted in most commercial half-runners and created much unhappiness among traditional half-runner enthusiasts, who want their beans to be both full and tender. At this time, many people are trying to locate and save the traditional half-runners, which have never been "improved" by being made tough for mechanical harvest.

Leather Britches—Leather britches, also called shucky beans, shuck beans, and in some areas fodder beans, are made from full green beans that have been strung, broken into pieces, and then dried. Traditionally they are dried by running a needle and thread through each piece and hanging them up in long strings behind a wood cookstove to dry out as quickly as possible. They can also be dried by spreading them out in a greenhouse on bedsheets, newspapers, or window screens. Still another way of drying them is putting them on window screens on a tin roof and bringing them in at night, or even putting them in a junk car with windows rolled up on sunny days. In the past eaten almost every day during winter and spring, they are now served mostly on special days such as family reunions, weddings, anniversaries, Thanksgiving, Christmas, New Year's Day, and other holidays. Drying beans, the oldest way of preserving them, is still very effective. Properly dried and cooked, they are delicious.

Pink Tip Beans—There are many varieties of pink tip beans. The term *pink tip* refers to the tip of the bean becoming pink in color as the bean becomes full. The tip becoming pink also indicates that the bean is ready to be picked for eating fresh, canning, or making leather britches. Seeds of pink tip varieties can be white, black, brown, tan, striped, mottled, or speckled, depending on variety.

Bill Best holding leather britches. *Photo by Irmgard Best*

Pole Beans—These are the same as cornfield beans. When some gardeners stopped growing corn in their gardens, poles often substituted for cornstalks. They are often used in teepee style to give stability. More recently, poles have given way to trellises, which provide more room and more sunshine to the bean vines. They can also be made stronger to survive better in windy weather.

Seed Colors—Heirloom bean seeds come in many colors: They can be solid red, white, black, blue, purple, brown, tan, pink, beige, and other colors. They can be speckled, with any number of colors of specks on contrasting background colors. They can be striped, with the stripes being of many colors, and also have many background colors, such as black on white or dark brown on tan. And they can be mottled, which is when there is a combination of specks, stripes, and smudges. Some are solid in color except for the eyes, which are of a different color. Others are mostly of one color with smudges of another color in random patterns, all within the same hull. One bean with striking colors is the Turkey Craw Bean, which is tan on one end, buff on the other, and speckled in between. Some people are so enthusiastic about colored beans that they use them to create jewelry.

Seed Shapes—Seed shapes can range from nearly perfectly round to oblong and nearly flat. Most are oval and elongated. Cut-short varieties have varied shapes, with beans within a single hull being of many shapes because of the pressure of the seeds against one another within the hull during the growth period.

Shelly Beans—This term refers to a bean shelled from a mature full bean before the hull and seed dry out. The beans are then cooked without the need for rehydration as would be the case with dry beans. Beans also can be frozen at the shelly stage and cooked later as one would cook the freshly shelled beans. Shelly beans are very popular with many old-time gardeners and others who knew them as children.

Snap Beans—At one time, almost any bean picked green for eating fresh or drying would be called a snap bean, because after being strung they would snap or break quickly and cleanly. With most commercial beans now having been bred to be tough and stringless to withstand mechanical harvesting without breaking, the term *snap bean* is rarely used, because the modern bean does not snap or break cleanly. This is also why so many are now canned and cooked as whole beans before the seed begins to develop. However, most heirloom beans picked at the green stage, even when full, are still snap beans.

Soldier Beans—Having a crop of beans in which the beans line up on the bean stem in formation is a mark of a good bean crop. When the beans are lined up one by one or two by two until the stem contains six to twelve beans, picking them is easy and they can be picked a handful at a time. Such beans are sometimes called soldier beans. The number of beans per stem is often limited by weather conditions, and too much heat will result in a lot of beans dropping off shortly after the bloom stage. Individual vines under good conditions may have one hundred or more bean pods containing seven to eight hundred seeds at the same time.

String Beans—Most heirloom beans are string beans. This means that they have at least one string per side, while some have two on the inner section (one on each half, making three strings altogether but still easily removed). These strings have to be removed prior to cooking or drying. Exceptions to this rule are some varieties of October/fall beans. Many people, especially those who use beans as a principal part of their diet, will not plant or purchase beans that do not have strings, for they consider stringless beans to be of poor quality, both in texture and in flavor.

Stringless Beans—Most beans produced by modern plant breeding and most sold in commercial catalogues are stringless beans. The downside of stringlessness is toughness: this is why commercial bean customers are advised, "Pick while young and tender" or "Don't let lumps (seeds) appear in your beans" or "Our beans are grass-like." Stringless beans typically have to be harvested before the protein (seed) appears.

Variants—Beans often mutate or cross and then grow back true to the new form. *Variant* is a word often used to characterize the new bean. Most such beans are then sold or distributed as a variant of the bean from which they mutated. However, within a few years, such beans usually assume their own names and have identities completely separate from the original parent bean. In the meantime, variants might even have their own variants. This is why there are so many heirloom bean varieties.

Wax Beans—This type of bean, usually yellow or light colored in appearance, has a hull somewhat thicker than other cornfield beans and a waxy feel to the touch. It is often used in three-bean salads. This is a popular heirloom bean in some areas, not very well known in others, and virtually unknown in still other areas.

As heirloom beans continue to make a comeback in gardens, restaurants, and the marketplace, identifying them by type will again be necessary, keeping in mind that an individual variety can be a combination of types. For example, a variety can be a long speckled greasy cut-short cornfield bean or a large, multicolored, bush butter bean.

Those who have never eaten heirloom beans are in for a treat, as more and more people are finding out, especially as heirloom beans are becoming more available at farmers' markets and as more people are growing them in their gardens.

Tomatoes

UNLIKE BEANS, tomatoes were not particularly prominent historically in the diet of Southern Appalachian people. Harriette Arnow, in her splendid book *Seedtime on the Cumberland,* discusses beans on 15 pages of her 449-page book. She goes into great detail about growing beans, harvesting them, cooking them, and preserving them in various ways. She discusses how much they were worth with respect to other foods commonly grown by the settlers and those who developed the Cumberland River area of Kentucky and Tennessee. But in this widely respected and thoroughly researched book about the early history of a large part of the Southern Appalachians, she never uses the word *tomato.*

However, tomatoes came to be highly regarded as a staple in the diet of most mountaineers beginning in the twentieth century. While the numbers of varieties are far fewer than those of beans, they have become increasingly prized for their flavors, textures, colors, and sizes. Like bean varieties, tomato varieties fall into distinctive types. And as canning became available, certain varieties became known for their suitability for canning purposes.

Heirloom Tomatoes

One of the earliest memories of my childhood has to do with sitting on the front porch with my grandfather on rare warm evenings in February. As dusk approached, we would notice a mountainside about a mile north of our house as it seemed to be burning. This was not a forest fire, however; it was Manson McElroy burning his tobacco beds on the mountainside above his house, usually four or five beds, each a hundred feet long.

Tobacco bed burning was a regular ritual of every farm family in our community. The beds were burned to kill weed seeds and other organisms that might interfere with the growth of young tobacco plants. After the beds cooled down enough and had been carefully raked, the tiny tobacco seeds would be mixed in a ten-quart bucket of fertilizer and sowed by hand, spread evenly with the fertilizer, over the tobacco beds.

Cotton tobacco canvas was then carefully placed over each bed to speed up germination and protect the young tobacco plants from frost. Should a frost kill the young plants despite the precautions, the process had to be done over—minus the burning. The second seeding was accompanied by fresh horse manure under the soil to generate heat to speed up germination. The second beds, called "hot" beds, usually worked. We needed to prepare hot beds only a couple of times during my childhood.

For many, if not most, farm families, the tobacco beds did double or triple duty. At the ends or along the sides, most people planted tomato seeds, and some planted both tomato and pepper seeds. The tomato and pepper plants grew at about the same speed as the tobacco plants and often were planted in separate rows

Bill Best with heirloom tomatoes at Lexington Farmers Market. *Photo by Michael Best*

alongside the tobacco rows, usually on the upper side of the field but sometimes on the bottom, depending on the fertility of the soil.

Few people at that time pruned or staked their tomato plants; instead, most let them sprawl on the ground. This resulted in many of the tomatoes rotting or sunburning, but there were generally enough to eat fresh and to make tomato juice or can as well. Additionally, the tomatoes typically ripened just when tobacco was becoming mature. Many farmers and their spouses and children would take saltshakers with them when topping, suckering, or cutting tobacco and were thus able to refresh themselves with tomatoes picked directly from the vine, salted to taste, and eaten whole, providing a good source of liquid at the same time. The tomatoes seemed especially tasty when accompanied by the smell of tobacco.

I noticed that tobacco and tomatoes seemed to do very well together, but not until years later did I realize that they are both from the nightshade family and obviously require similar soil nutrients and growing conditions. Other edible plants from the nightshade family include peppers, potatoes, and eggplants, all of which contain nicotine, just as tobacco and tomatoes do.

A few years ago at the Lexington Farmers Market, I had a regular customer who started buying about three times her usual amount of tomatoes. One day she remarked that she seemed to be developing an addiction to tomatoes. I casually asked her if she was trying to give up smoking at that time, and she proudly pointed to the patch on her arm and said that she was in her fourth week without cigarettes. When I pointed out that tomatoes also contained nicotine, it suddenly dawned on her that she was still getting a little of her fix of nicotine but in a much more benign form. She continued buying a lot of my tomatoes until my season ended.

Most families in my home community grew tomatoes in their tobacco patches and in their gardens as well. These were heirloom varieties, of course, but we did not call them that: they were just tomatoes of different colors, shapes, sizes, and flavors. They ranged from the very large beefsteak varieties to the much smaller ones we called "Tommy Toes."

Tommy Toes never had to be seeded in a tobacco or other plant bed, because there were always enough "volunteer" plants that it was necessary to thin them instead of planting them. Most of our Tommy Toe plants grew up around the hog pen, where we started fattening hogs for butchering in late November. The hogs were fed slop, which contained all vegetable matter left over from our meals, while the dogs were given any leftover meat. As the Tommy Toe seeds passed through the hogs, they became dormant until the following spring, at which time they came back to life as new plants and the cycle began again.

I do not even remember the names of the tomatoes we grew, just that all the flavors were good. I took tomato sandwiches with mayonnaise and salt to school for lunch as long as our tomatoes lasted, even if there was no light bread available and I had to take them on biscuits instead.

When the larger tomatoes gave out in the fall, there were usually still plenty of Tommy Toes for another two weeks, so I took those on sandwiches instead. My most memorable fall season as a child was one in which we did not have a killing frost until after Thanksgiving, at which time I picked tomatoes from under freshly fallen snow for Thanksgiving dinner. Such was my love of tomatoes that when the season finally ended, I took mayonnaise sandwiches for my lunch for two more weeks and pretended that I was eating tomato sandwiches.

A Tomato Tale

I did not have my own garden until I was twenty-seven years old, when my wife and I moved to Berea, Kentucky, to work at Berea College. We quickly bought a farm and had our first garden on land that had been cultivated for decades, and the previous owner had taken good care of the garden spot. We found the soil to be very rich and easy to work.

Until I was twenty-six years old and through college, graduate school, and the army reserve, I had lived at home during the summers and did not concern myself with garden vegetable varieties. My mother always saved seed, and her seed stocks sufficed for our family gardens.

That stopped after we moved to Berea, and I started buying seeds and plants from local suppliers. Although I did not think the tomatoes were as good as the ones I had grown up with, I naively attributed that to the differences in soil and not to differences in varieties. (I was not aware that tomatoes were being genetically manipulated to make them ship and pack more easily and stay on the shelf much longer than before.)

We planted far more than we needed and gave away a lot, but after we had been gardening for two years, I decided to grow several hundred tomato plants to see whether I could sell them commercially to local groceries and grocery suppliers. This was in 1965. Farmers in my home county in North Carolina were beginning to grow tomatoes for the commercial market through marketing cooperatives, and my family had a long tradition of growing, trading, and selling vegetables, digging and selling wild ginseng roots, and trapping fur-bearing animals to sell their furs during the winter. Not knowing whether I might have enough tomatoes to bother trying to sell commercially, I made no marketing plans and decided to market by the "seat of my pants" should I have enough.

We had leased a small acreage of high ground a mile from our farm, bought plants of the Rutgers variety, and started our venture. Almost everyone else's early gardens were killed by a late frost, but because our plot was on high ground, our tomatoes were spared and we had ripe tomatoes three weeks before most other gardeners in the area.

Continuing my experiment, I went to Lexington to talk with the produce manager of the Kentucky Food Store's warehouse, a company that went out of business a few years later. I was familiar with the company, because it supplied the local Berea grocery where we bought our weekly groceries. The produce manager tasted our tomatoes, said they were of high quality, and said that he would take all I could bring him.

About three days later, I loaded several bushels into our station wagon and went back to see him. When he saw my tomatoes, his mood changed dramatically. He called me ignorant and naïve and wondered aloud why he had wasted his time talking with me. I asked him what the problem was, since the tomatoes I brought were just like the ones he had liked so much a few days earlier.

He informed me that he wanted green tomatoes to be kept in storage until one of his stores wanted tomatoes. He would then "gas" them (his phrase) en route to their destinations to give them color. Even then, they would be expected to have a shelf life of weeks, and the final product, I found out later, would be expected to have very little flavor.

Of course, the modern tomato is much tougher and longer lasting than the ones I was working with more than forty years ago, having practically no flavor and the texture of Styrofoam. Modern tomatoes are also expected to have a shelf life of thirty-five days after being taken out of storage, gassed, and put on the store shelves.

One of my experiences a few years ago tells the story concisely and completely: About ten years ago, my wife and I wished to have a tomato to go with some okra we had frozen the previous summer. I bought a large red tomato at our local megamart, and we put it on a shelf in the kitchen to continue ripening and soften a little to indicate that it was ready to eat. About two weeks went by and it still had not softened, so we decided to try to eat it anyhow. After the okra was fried and spread on our plates, awaiting the tomato slices on top, we discovered that the tomato slices had a smell akin to formaldehyde and a very disagreeable flavor. Not wanting to completely waste the tomato, I decided to take it to our son's chickens about a hundred yards down the creek from where we live.

The chickens ran to the tomato slices just as they did to leftover tomato slices when we fed them vegetable scraps during the summer. After one sniff, however, they turned and went quickly the opposite way. Two weeks later the tomato slices were still lying on the ground, seemingly indestructible.

Technology had triumphed. The tomato had probably been picked some months earlier in Florida, Mexico, or California. More than likely it had spent some weeks in cold storage and had then been gassed to add color on its way from a warehouse to its final megamart destination, perhaps two thousand or more miles away. After all that time, there was no way to make it have flavor or a palatable texture. It had color and water and some unappetizing chemicals under its skin. But even the chickens would not eat it.

Tomato Flavors

Plant scientists have identified at least thirty-four acids and sugars that give tomatoes their distinctive flavors. Volatile compounds are also thought to influence flavors, with the sense of smell contributing to what a tomato tastes like.

Acids are more prevalent in red tomatoes, which has led to a tradition of using primarily red tomatoes for canning purposes and for making tomato juice. Acidic tomatoes also will liven up a tossed salad, and many people prefer highly acidic tomatoes for their tomato-and-mayonnaise sandwiches.

Pink tomatoes are known for being high in acids and high in sugars as well, which gives them a sweet and tart taste simultaneously. Many Appalachian heirloom tomato varieties are large pink tomatoes; one slice might be as large as a small plate, covering a slice of bread with tomato sticking out on all sides. Individuals who have chosen to work with one tomato for decades typically have moved toward a pink tomato. The Tomato Grower's Supply Company in Fort Myers, Florida, carries several heirloom pink tomatoes from Appalachian states, including the Mortgage Lifter and Richardson, among others.

Pink tomatoes often dominate tomato sales at farmers' markets, especially farmers' markets established decades ago, and are becoming increasingly sought after by chefs. They are chosen by those who simply want to cut up a tomato for a salad (or dessert) or sit down with a tomato, salt, and sometimes pepper, for a snack or an entire meal. Being high in both acids and sugars, pink tomatoes have what older people seem to regard as "old-fashioned" flavor.

When enticed to buy pink tomatoes by being offered sample slices at farmers' markets, customers will often buy at least one to take home with them in addition to red tomatoes. Frequently they come back the next time wanting only pink tomatoes, and they often learn the names of pink varieties to know what varieties to ask for. Because there are very few commercial pink varieties left, customers usually have to pick among heirloom varieties.

Yellow tomatoes are usually higher in sugars than are tomatoes of any other color, with the yellow German varieties considered the sweetest of all large varieties. The many yellow German varieties begin to ripen yellow and develop red stripes as they continue ripening, often developing a deep red blush in the center of the tomato. Yellow German tomatoes are widely thought to be Amish or Mennonite in origin, but variants of them are grown throughout the Southern Appalachians.

While red, pink, and yellow tomatoes are the dominant types of tomatoes, there are many other colors as well. Some are a deep orange, some are still green even after ripening, others are brown, some have a deep purple hue, and some are almost black. All of these colored tomatoes have their own distinctive flavors, and with so many acids, sugars, and volatile compounds involved, the list of flavors is almost endless. Individuals who have eaten only gassed tomatoes purchased in grocery stores have missed out on most of those flavors.

Appalachian Varieties of Tomatoes

Radiator Charlie's Mortgage Lifter

There is little doubt that the best-known Appalachian heirloom tomato variety is the Mortgage Lifter, more properly called Radiator Charlie's Mortgage Lifter. It was developed in the 1940s by M. C. Byles of Logan, West Virginia, where he was an auto mechanic.

Lacking in formal education but not in ingenuity, Radiator Charlie ran a radiator repair shop at the foot of a steep mountain. When coal and timber trucks or other heavily loaded trucks blew a radiator while going up the mountain, the truck drivers would have to back down the mountain to his shop, where he could fix it for them. He was also good with other mechanical tasks and, while a young man, flew small planes to deliver airmail. He also invented a new type of garden tiller but neglected to get a patent on his invention.

Starting with four varieties of tomatoes, Byles set out to create a tomato that would combine the best characteristics of those tomatoes into one. Placing a German Johnson variety in the center of a ring, he planted plants of three other large tomato varieties around it. He then used a baby's ear syringe to transfer pollen from the outer tomatoes to the one in the center.

After seven years of work on his tomato, he felt that he had stabilized it enough to start selling plants to other gardeners. Selling them at a dollar each when the going rate for much labor was a dollar a day proved that he was a good salesman as well as a tomato breeder. Part of the lore of Radiator Charlie's tomato is that it was so popular that he was able to sell enough plants at a dollar each to pay off the mortgage on his house in six years.

A large pink tomato with a small blossom-end scar and not as prone to cracking as many other varieties, Radiator Charlie's Mortgage Lifter is still highly regarded among serious gardeners and is becoming increasingly popular with chefs who frequent farmers' markets. The tomatoes often weigh a pound or more and have an excellent blend of sugars and acids.

(Part of the above information is from an interview of M. C. Byles when he was eighty-five years old by his grandson, Ed Martin, of Virginia. Martin was interviewed by Jeff Young and Jeff McCormack for *Living on Earth.* Radiator Charlie lived to be ninety-seven years old.)

The Vinson Watts Tomato

While Radiator Charlie's prize tomato is the product of a deliberate cross of four varieties of tomatoes in the 1940s, the Vinson

Vinson Watts Tomatoes, grown by Jeri Rumsey. *Photo by Jeri Rumsey*

Watts Tomato is from fifty-two years of deliberate improvement of a single tomato variety.

Vinson Watts, who lived in Morehead, Kentucky, from 1967 until his death on March 17, 2008, was born and raised in Breathitt County, Kentucky. He received his college education at Berea College in Kentucky and went back later, in 1956, to become the associate dean of labor at Berea, working under Wilson Evans, the dean of labor at that time.

Evans and Watts were avid gardeners and often talked about their tomatoes. Evans usually planted a large pink tomato that his family had maintained for generations in Lee County, Virginia, where he grew up. While he was pleased with his family tomato, he also wanted to branch out and try some of the hybrids that were becoming popular in the 1950s. Evans asked Watts, who had already been growing the Evans tomato, whether he would be willing to keep the tomato pure while Evans tried some other varieties.

Vinson Watts agreed and was so sold on the idea of working with that one tomato that he grew only that variety in his gardens from 1956 until his death in 2008. He did experiment with other tomatoes in the gardens of other people but decided to keep the first one pure in his own garden and just select seeds from it.

From 1956 through 1967, Vinson Watts worked for Berea College and gardened in Berea, Kentucky, and then he went to Morehead State University in Morehead, Kentucky, to become their first personnel director. There he started gardening more intensively than ever.

Selecting each year for disease resistance, flavor, texture, and size, he gradually developed a very disease-resistant and increasingly flavorful tomato that, more often than not, exceeded a pound in size. Deep pink in color, it is increasingly being sought after by farmers' market customers, chefs in particular. Regular gardeners are now also buying seeds to grow their own.

About thirty-five years ago, Wilson Evans gave me some seed of his family tomato, which I grew along with many other varieties of heirloom tomatoes I was growing at that time. I was impressed by its flavor and saved seeds to try again in the future, but I chose not to grow it again, because it seemed too prone to disease with too many cracks around the top (vine end) of the tomato. Consequently, I am quite aware of the improvements made in the tomato over a more than fifty-year period under the stewardship and loving care of Vinson Watts.

By selecting only the most flavorful, well-formed, strong, and disease-resistant tomatoes for saving seeds each year, Vinson Watts achieved far more than plant breeders who produce hybrid tomatoes during the space of two or three years, usually focusing only on production and shipping qualities. He did not have to compromise flavor and disease resistance, since he was not trying to produce shipping-type tomatoes.

I have grown his tomatoes side by side with very good hybrid tomatoes in high tunnels and have found them to be the most disease resistant of all the tomatoes I currently grow, more than 200

Vinson A. Watts (*left*), presented with lifetime achievement award for his tomato, and Bill Best (*right*). *Photo by Patty Watts*

varieties in all, and more than 450 varieties during the past forty-nine years. (I still grow some of the best available hybrids but mostly stick with heirlooms.) It is even resistant to botrytis, an especially troublesome fungal blight in greenhouses and high tunnels, the first tomato I have seen with such resistance.

The *Lexington Herald-Leader* in Lexington, Kentucky, featured Vinson Watts and his tomato in a long article in the summer of 2006. The article was quickly reprinted by many other newspapers across the country, and he was soon deluged with inquiries about the tomato. I also started featuring his tomato on our website, www.heirlooms.org, and started receiving many orders for seeds of the Watts Tomato.

Chefs buying his tomatoes from me at the Lexington Farmers Market started featuring his tomato, by name, in some of their dishes, and the tomato variety became increasingly popular. Watts has also been featured by other publications, including *Cooking Light* magazine in August 2007.

While modern technology has been a mixed blessing for food production and more often than not proves quite negative for vegetables, the Internet has been a boon for those interested in locating and growing heirloom beans, tomatoes, and other vegetables. It has certainly allowed Appalachian people to continue family and cultural connections that are too often erased by the passage of time. Information given to me by those who knew of and participated in the development of the Vinson Watts Tomato is an excellent case in point.

During the fifty-two summers that Vinson Watts worked with his tomato, I would see him occasionally at different Berea functions and professional meetings in Kentucky. We always discussed gardening, and our conversations focused on gardening more than any other item we could have discussed. I knew of his experimentation with the tomato, but I did not know how thoroughly he was involving his neighbors on Allen Avenue in Morehead until people who knew him started contacting me over the Internet. One such connection happened on March 7, 2010, when I received the following e-mail from Barbara Thomas.

> I checked out your catalog and saw something that was to me unbelievable. Your Vince Watts tomato seed. My maiden name is Allen. I grew up on Allen Avenue in Morehead, Kentucky. Allen Avenue was named after my Grandfather who owned all the land there and ran a slaughterhouse on it. Vince Watts also lived on Allen Avenue and grew and bred those tomatoes in my backyard. Although I am 48 years old I grew up "trying" out every year's new experimental tomato. He topped himself every year but the last few as he found that they could not be improved upon. He gave me seeds every year to grow for myself. I had a little bad luck one year and had a house fire that ruined my seeds. That was the year that Vince got sick. I figured that my good tomatoes were

> gone for good. Now you know the story. I could not believe my eyes when I saw his seeds in your catalog. I will send my order to you ASAP and those tomato seeds will surely be included.

I answered her letter and asked for more information about how he involved neighbors with his seed experimentation and saving. Because a person's spending so many years working with one plant is so rare, I thought it important to get his process down from a person who was involved with it from the time she was a child. She promptly wrote back at 7:22 a.m. the following day, March 8, 2010, with this additional information:

> Most of those that I know that "helped" in his quest have died. Grace Lewis and Harold Bellamy I know grew the tomatoes. Lisa, Harold's daughter, is one of my best friends and will know much more than I about them so I will get in touch with her sometime today. I also know that Vince grew tomatoes in the garden behind my house and also had what I call an experimental type garden in and around his home. He gave people seeds and plants so that he would have information on the plants' hardiness and blight resistance in other terrain and the amount that the plant produced. I knew when we grew a few in pots he always wanted a couple of them. At first we had no idea what he was up to. We later found out because we would save some seed to grow the following year and he would insist on us growing the "new" plant instead, or at least add the new one to the other plants we were growing. When I find more information I will let you know.

Almost two years after I had this e-mail exchange with Barbara Thomas, in January 2012 I met a lady at an agricultural conference at Morehead State University who said that she still maintained the Vinson Watts Tomato from seeds he had given her when her high school–age son worked with him during the years prior to this death.

Claude Brown's Yellow Giant

Claude Brown, a native of Pike County, Kentucky, was an educator for most of his life. After reading about our farm and gardening operation in the *Rural Kentuckian* in 1988, he wrote a letter to me about his experiences with gardening. A series of letters led to his sending me seeds of a deep yellow tomato that he had worked on for many years. He called it Brown's Yellow Giant.

I grew his tomato and found that it lived up to the description he had given for it. When I started to sell seeds of his tomato a few years ago, I found that Brown's Yellow Giant was already being sold on the Internet by another seller of heirloom seeds. To avoid confusion, I called the tomato I was selling Claude Brown's Yellow Giant (although I do not know whether the Brown's Yellow Giant available through this other seller is the same).

Like Vinson Watts, Claude Brown worked on his tomato for decades, bringing about as many improvements as possible. Claude Brown's Yellow Giant is certainly a notable tomato, as is true in the few other instances in which a single individual worked for many years on improving one variety. It is one of the best-selling seeds on our website and also is a good seller at the farmer's markets we attend.

Other Tomato Developers

Over the years I have found two other people who have worked for a long time on developing particular tomatoes, and they deserve credit as well for work well done.

One is Zeke Dishman of Windy in Wayne County, Kentucky. He has worked with a large red tomato that is quite tasty and has few seeds. He originally received the tomato from a lady on whose land he was cutting timber, but he adopted the tomato as his own and worked for decades to perfect it to his satisfaction.

(Above)
Zeke Dishman's
tomatoes

(Left)
Zeke Dishman

Willard Wynn's yellow German tomato

Another is Willard Wynn of Rockcastle County in Kentucky. Originally from Harlan County in Kentucky, Willard worked with his yellow German tomato for decades. I have grown it for over twenty years, and it is consistently a good producer. As with other yellow German types, deep yellow with red streaks, it has a very sweet flavor.

Saving Tomato Seeds

Compared to saving bean seeds, saving tomato seeds is somewhat more difficult—much more difficult, many would say. This is probably the reason why there are far fewer family tomatoes than family beans.

In early times, most people just waited for a tomato to become overripe and then spooned out the seeds on a piece of cloth or a piece of paper (if available). When dried, the seeds would be scraped off with a sharp object such as a table knife, wrapped in cloth or paper to keep them dry, and kept in a dry place.

Knowledge of best seed-saving practices is now widespread, and most seed savers now quarter the tomatoes and mash them up a little, then let the seeds ferment in a container for several days. The pulp and seeds are stirred once or twice daily as the mixture ferments, until the seeds loosen from the pulp and start falling to the bottom of the container. The fermentation process is also helpful in killing organisms of some seed-borne diseases.

Water is then poured into the container, allowing the pulp to rise to the surface and fall over the edges. This will include some tomato seeds, but the seeds that float are usually not fertile and not worth saving. The seeds that remain on the bottom of the container after the pulp has been poured off are ready to save.

It is then time to pour most of the water out of the container, being careful not to let the seeds on the bottom move upward and spill over as well. I have found it best to pour the seeds and remaining water into a closely woven kitchen strainer, allow the remaining water to drip through, and put the still-wet seeds on a piece of waxed paper or a paper plate with a slick finish. This can be done by bumping the strainer on one's hand with the waxed paper or paper plate underneath. It is then time to separate the seeds as much as possible from one another with a fork or other sharp-edged tool prior to drying, being careful not to cut into the still-wet seeds. Putting the wet seeds on a paper towel or a sheet of newspaper makes separating them harder, although some people who are just saving seeds for themselves will plant the seeds—towel and all—in their germination medium.

I put my seeds on waxed paper on a flat surface and let a ceiling fan on low speed dry them for one or two days, depending on the humidity of the room being used. In my house, one to two days is sufficient to dry them. I then scrape them off, separate any that have clumped together by twisting them in my fingertips, and let them continue to dry loosely another day or so. After they are dry, I seal them inside tightly sealed plastic bags, pushing as much air out as possible.

I used to keep my tomato seeds in a freezer, but for the past few years I have just tried to keep them in a cool dry area. Keeping them unfrozen in a refrigerator is probably best, but my method of storage yields good germination for up to twenty years. It is very important not to let the seeds get overly hot.

Apples

In the part of Haywood County, North Carolina, where I was born and raised, we had our favorite apple varieties, which included the June Apple, Northern Spy, Yellow Transparent, Winter Banana, and Horse Apple. We also had numerous sweet apples that did not have names and were probably grown from seed and not grafted. Still, the nameless varieties were good for drying and cooking, and some for eating fresh as well.

The June Apple, a red apple darker on one side than on the other, was the earliest and smallest, and we always ate them fresh. The Yellow Transparent was known for its sweetness, but it had to be eaten within a day or so of ripening, so the apples had to be watched closely. The Horse Apple was my mother's favorite for making green-apple pies, which she made in abundance during the two or three weeks before they ripened. The ripened Horse Apple did not quite have the flavor of other apples but was still good. The

Winter Banana was a good keeping apple but not really that flavorful. The Northern Spy had a distinctive flavor that made it our favorite for both eating and cooking. We also had an unnamed variety that was best for making dried apples for later preparation as fried apple pies. My personal favorite apple for eating fresh was the Sheepnose, a reddish apple with an elongated shape, similar to a sheep's nose.

Apples and other tree fruits have a long history in the Southern Appalachians, and on many farms, remnants of old apple orchards can still be found. During my several summers of measuring tobacco acreage for the federal government in Haywood and Madison Counties during the 1950s, I noticed that most small farms still had their own apple trees, sometimes suffering from neglect but yielding enough to meet the family's needs for fresh apples and for canning and making jelly, applesauce, and dried apples.

Historically, almost every extended family had at least one highly skilled tree grafter who kept favorite varieties going. Over time, and with the advent of so many fruit tree companies springing up, grafting skills have largely died out. The person in the Southern Appalachians who has done the most to rescue older apple varieties is Harold Jerrell of Rose Hill in Lee County, Virginia. Not only does he graft hundreds of heritage varieties, sometimes called "antique" or "heirloom" varieties, but he has also trained hundreds of other people in the art of tree grafting during the past twenty-five years, not only in Lee County but also in surrounding counties in Virginia, Kentucky, and Tennessee.

Recently retired, Harold was the longtime agricultural extension agent for Lee County, Virginia, a narrow county stretching fifty miles on the south side of Pine Mountain opposite Harlan and Bell Counties in eastern Kentucky. It joins Cumberland Gap where Virginia, Tennessee, and Kentucky meet and contains many miles of the Wilderness Road that Daniel Boone traveled and, to some extent, created while moving back and forth between North Carolina and Kentucky. The following is the story, in Harold Jerrell's words, of his work in his home county.

Harold Jerrell

Grafting Fruit Trees

HAROLD JERRELL

As a young teenager in the mid-1960s, I remember visiting our neighbor's farm and picking up apples from an old abandoned orchard. Our neighbors lived in another state and gave my parents permission to harvest apples from this orchard *that* had been planted several decades earlier. Part of a limestone chimney stood next to the orchard and was the only remaining clue that a family had lived next to the orchard many years ago. There were fewer than a dozen trees living, and the only variety I remember was Summer Rambo.

Summer Rambo was the apple I remember most from my youth. The flat, green fruit with red stripes on the side facing the sun would be washed, peeled, and cut into slices about three-eighths of an inch thick. The apple slices would be spread out over a sheet and placed on the tin roof of a building during the day to dry. In late afternoon, the apples would be brought back into the house to avoid picking up moisture from the cool night air. The next morning they would be taken back to the roof. This process would be repeated until the apples were dry and flexible like leather.

The dried fruit would then be placed in cloth sacks and, properly stored, would last several months. Throughout the year Mom would take out the necessary amount of dried apples, wash them, soak them in water, and cook them over low heat to form a thick paste. Sugar would be added to sweeten the cooked apples. She would then roll out dough to a desired thickness, place a generous amount of cooked apples on one half of the dough, and fold the other half over, sealing the edges by crimping the dough

Charles and Sarah Davis, ages ninety-nine and ninety-four, seed savers from Otto, North Carolina

together. The dough would be fried in an iron skillet at just the right temperature to create a golden brown fried apple pie. Mom will soon be eighty-three and can still fry pies as delicious as she did years ago.

The farm where I grew up and live today was at one time part of Silver Leaf Nurseries. During the 1770s, Daniel Boone had blazed a trail through this area that would become known as "Boone's Path" as he made his way to Cumberland Gap and the soon-to-be state of Kentucky. The name *Silver Leaf* was supposedly given to the community by Boone as he sat down one day to rest at a spring. Silver maple shadows danced across the surface of the water, and Boone called the spring "Silver Spring." Later the name *Silver Leaf* would be chosen for the community, and a school and church would be built, providing for the education of the mind and the soul. Although the school no longer exists, the community of Silver Leaf and Silver Leaf Baptist Church still exist today.

The C. C. Davis family started Silver Leaf Nurseries in 1776. Salesmen carried full-colored catalogs as they visited potential buyers. Although several generations have passed, descendants of the Davis family still manage a local nursery business.

I grew up on a farm and loved nature. I would spend much of my spare time walking the hills and valleys with my dogs. During high school I became a member of the Future Farmers of America (FFA) and was selected to be on the forestry judging team. However, it was an event in 4-H that led me to choose forestry as a college major. When I was in the seventh grade I entered an exhibit in the forestry category. I won first place, and from that moment forward, my goal was to become a forest ranger.

After graduating from college with a degree in forest management, I worked ten years for the Virginia Department of Forestry. I then transferred positions to become the Agricultural Extension Agent for Lee County, Virginia. One of the programs I chose to teach was apple grafting. The response to learning the technique was overwhelming. Although I have taught grafting for twenty-five years, the number of individuals that continue to participate annually surprises me. Some travel miles to learn the technique of grafting. I think many attend not only to graft a new tree but also to renew fellowship with other apple enthusiasts.

I am of the opinion that the interest and success of this project has been the desire to bring back what many call "antique" or "heirloom" varieties. Variety names such as Benham, Mountain Boomer, Strawberry, Liveland/Lowland Raspberry, Arkansas Black, Pound Sweet, Sheepnose, Horse Apple, Early Harvest, Spice Apple, Johnson's Fine Winter, Limbertwig, Winter Banana, and Smokehouse are bound to catch one's attention or bring back memories. Fifteen varieties of Limbertwigs were found in the Smoky Mountains when the land was being purchased for what would become the Smoky Mountains National Park. Varieties such as Royal Limbertwig, Brushy Mountain Limbertwig, Swiss Limbertwig, Sweet Victoria Limbertwig, Smoky Mountain Red

Limbertwig, and my favorite, Myers Royal Limbertwig, are just a few examples.

As I travel throughout Lee County I have discovered a few Limbertwig apple trees still bearing in some old abandoned orchards. Older individuals have shared with me that Limbertwig varieties were grown through the county and were well known for their ability to "keep all winter." The origin of the Limbertwig apple is not known, but evidence points to the mountains of East Tennessee. One of the interesting characteristics of the Limbertwig varieties is that they often have drooping or weeping branches.

In addition, the Limbertwig apples are said to have a distinct taste, different from other varieties. Several years ago I conducted an apple tasting contest where sixty varieties were cut into small pieces. Approximately twenty-five people judged the fruit based on taste. The apple chosen as "having the best flavor" was a Myers Royal Limbertwig. I have this tree in my orchard and would have to agree that it is a good apple. Not only does it have a distinct flavor, but the tree is also disease free. Another positive characteristic of this variety is that the blossoms have good frost tolerance.

Frequently individuals call and ask which varieties will grow well in southwestern Virginia. Some varieties that I recommend are as follows.

Early Summer Apples—Yellow Transparent, Early Harvest, Lodi, and Red June

With the exception of Red June, all of these apples are yellow skinned and tart and ripen in early July. The apples make excellent applesauce. *Lodi* is a vigorous growing tree that produces a large apple but has a tendency to bear fruit biannually. All of these apples bruise very easily, and if they are allowed to fall from the tree, a large portion of each apple will be damaged. They do not store well. *Red June* ripens about the same time as

Yellow Transparent, Early Harvest, and Lodi but differs in several ways. The skin is medium to dark red, the apple size is small, the apple is conical in shape, and the fruit is not as tart as the other early varieties.

Mid–Late Summer Apples—Benham, Summer Rambo, Wolf River, Spice, Mollie's Delicious, Strawberry, Chango Strawberry, and Ozark Gold

The most sought-after variety is *Benham.* Benham was listed in local nursery catalogs in the late 1870s and may have originated in East Tennessee, maybe Claiborne County. The medium-sized fruit is yellow in color. The side facing the sun will often have a pinkish blush. Benham bears annually and ripens in late July. Benham is also very susceptible to fire blight and cedar-apple rust. When "old-timers" are asked why they grew this variety, since it does not have good disease resistance, the answer most often given is *flavor!* Benham can be dried, frozen, canned, or made into applesauce.

Summer Rambo is reported to have originated in France as early as 1535. The large flattened fruit is greenish yellow with red streaks and has a red blush on the sunny side. The subacid fruit can be used in many ways, such as eating out of hand, drying, or applesauce. Summer Rambo makes fine tasting apple butter. The tree is susceptible to cedar-apple rust but is resistant to many other diseases.

Wolf River produces a very large apple and has a dry flesh. The skin color is medium red. The apple is best used for drying and cooking.

Spice (may be called Old Virginia Spice in some locations) is a small apple with pale yellow skin. It is a reliable producer, subacid in taste, and disease resistant.

Mollie's Delicious is a large red-and-yellow-skinned apple with a sweet taste. It is slightly susceptible to scab. The blossoms have good frost tolerance, and the fruit stores well under refrigeration.

Strawberry and *Chenango Strawberry* have similar characteristics. Strawberry is earlier ripening and is a medium-sized conical-shaped apple. The skin color is yellow with red stripes. I am not aware of any apple that produces as sweet a fragrance as the Strawberry. Both apples ripen over a long period of time and have a subacid taste. The Chenango Strawberry is larger than the Strawberry and ripens later. Both apples are subject to brown rot and fire blight. The apples can crack if left on the tree, and the sweet fragrance attracts both June bugs and Japanese beetles.

Ozark Gold is a Golden Delicious cross with a sweet flavor and produces annually. It is resistant to cedar-apple rust, unlike its parent, Golden Delicious. The fruit is large in size and has a yellow skin with a blush on the sunny side. The flavor is sweet, and the apple is good to eat fresh.

Late Apples—Red Delicious (Hawkeye), Virginia Beauty, Winesap, Myers Royal Limbertwig, and Arkansas Black

The original *Red Delicious* was called *Hawkeye* and does not resemble the Red Delicious we purchase in today's stores. Those who may still have a Hawkeye claim that the fruit is much better flavored than the many sports or mutations that have been developed from it. The skin color is yellow with red stripes, and the flesh is sweet. The apples store well.

Lee Calhoun Jr., in his book *Old Southern Apples*, states that "if Red Delicious had not come along at about the same time, I believe *Virginia Beauty* would have become the most important commercial apple in the South." Virginia Beauty produces a large dark-red-skinned apple that may have a considerable amount of russet on the skin around the stem. This feature is so prominent that it makes identification easy. The flesh is subacid and very juicy. I grafted and planted a Virginia Beauty several years ago. As I walk to the barn to feed my sheep, I pass this tree, and I often stop by for a moment to enjoy the flavor of a tree-ripened fruit.

Winesap probably has been planted more in the South than any other variety has. It is an excellent keeper and is disease resistant. It seems to bloom a little later than most varieties and often produces a crop when others have been killed by late spring frosts. The yellow-fleshed Winesap is medium to large in size with greenish-yellow skin that may be red striped and have a reddish blush on the side facing the sun. When harvested, the apple is very tart, but it mellows as it ages and is an excellent apple to eat during late winter.

Myers Royal Limbertwig is one of several Limbertwig varieties that came from the Smoky Mountains of East Tennessee. The fruit is dark red in color, very large, juicy, and aromatic. It ripens late and is excellent for eating fresh, for cooking, and for making apple butter. This variety requires little pruning and is disease resistant. It also produces apples when other trees are killed by late spring frosts.

Arkansas Black is a medium-sized apple with a very dark red to black skin. The flesh is yellow and the flavor is tart. Arkansas Black is very firm when harvested and stores very well. During my childhood, I often dug out these apples from beneath Kentucky Fescue grass. The light freezes just seemed to mellow the apple, and they sure tasted good.

There are many other varieties that I could have listed as being worthy of growing. However, the above are those that often match the desires of those wishing to start a small orchard. Most varieties have good disease resistance or have such a desired flavor that fruit growers are still willing to grow them. It is this author's opinion that many of the varieties sold commercially today are bred only for eye appeal. Until the development of Granny Smith, almost all apples sold commercially were either red or yellow in color. However, many backyard apple growers will agree that beauty is only skin deep, and often our "antique or heirloom" varieties many not have the prettiest skin color, but the flavor is out of this world.

Corn

ONE OF MY earliest memories is going with my father to the mill to have corn ground into cornmeal. The mill was several miles downstream from our house after several creeks had run together to make a stream large enough to turn the overshot wheel and grind corn. The elevation of the land at that point was suitable to locate the mill: the raceway where the water ran lost altitude quickly and thus did not have to be too long before the water turned the overshot wheel.

We would take enough corn to last us two or three weeks, and the mill owner would take his pay by keeping a portion of the meal produced. He would then sell it to other people who did not bring their own corn to be ground. No money was ever exchanged.

We always grew white corn to be ground into meal, for we, like most of the other people in our community, preferred to have corn bread made from white instead of yellow cornmeal. At that time, most of the corn grown locally was white open-pollinated corn because hybrid, mostly yellow, corn varieties had not come to our community.

When hybrid corn varieties did become available, my father was the first in the community to try them. I was in the Future Farmers of America (FFA) club at that time and in the 4-H club as well. My FFA projects were animal related, while my 4-H projects were field crops, specifically corn.

Daddy gave me an acre of land in the summer of 1951 on which to grow my 4-H corn project. I tried a just-released hybrid, U.S. 282, and dutifully kept all records on that acre. Instead of simply replanting missing hills of corn, I used a shovel and transplanted corn from rows where the stalks were too thick to areas where the stalks were too thin. This brought about some good-natured teasing from my parents, who asked me if I was trying to set a new corn record. They thought the transplanted cornstalks would not live, but most of them did, thanks to some timely rains, giving me an excellent stand.

At the end of a nearly ideal growing season, when my acre of corn was tested by the 4-H agent for approximate yield, the test revealed that my projected yield (a near record) was sufficient to justify a completely measured and weighed harvest. My father decided to go along, and we rounded up about twenty people, including uncles, cousins, and other community people, along with two agricultural agents. The acre was hand harvested and taken into town by trucks to be weighed.

To the surprise of a lot of people, my acre of corn set a new North Carolina per-acre production record of 163.19 bushels. I was fifteen years old. At first, the thinking of the state agricultural establishment was that I should be declared only the 4-H record holder, but after much discussion at several levels, it was decided that I should be declared the overall champion corn grower because my record beat the existing record of 148 bushels per acre by 15 bushels. It remained the record for nearly twenty years.

My acre of corn sent me to the National 4-H Congress in Chicago that November and allowed me to meet many other young people from many states who were interested in agriculture and rural communities in general. The trip also deepened my interest in agriculture, an interest that has remained with me throughout my life.

While my father continued growing hybrids for the rest of his life (he died in 1982 at age eighty-three), many farmers in our community refused to try them for a number of years, and some never became comfortable with them, choosing instead to remain with the open-pollinated varieties even though these varieties did not produce as many bushels per acre as did the hybrids. Those farmers were wise in ways even they did not realize, because the varieties thus saved are important today in maintaining genetic diversity.

The United States became so dependent on hybrid corns that in 1970, 15 percent of the American corn crop was devastated by Southern corn leaf blight. Corn was fast becoming a monoculture crop by that time, with much of the genetic diversity previously existing in corn having been bred out by commercial plant breeders; most farmers were becoming dependent on companies producing hybrid corns, which were increasingly susceptible to blights.

I no longer grow hybrid corns and instead am trying to grow open-pollinated varieties and work with others who are trying to save the open-pollinated varieties, which are likely to be increasingly important in the future.

There are still many people in the Southern Appalachians who are maintaining their personal or family open-pollinated varieties of corn and many others who are maintaining several varieties. One of the most important growers of heirloom corn varieties is Tony West, now living and gardening in southern Ohio.

Tony, with his Cherokee heritage, is particularly interested in collecting and maintaining varieties associated with the Cherokees. However, he is also one of the most knowledgeable collectors

Tony West

around when it comes to the history of corn and its long-term development. Tony states:

I have been growing things since I was a child (starting with maple trees) but started to garden in the late 1980s. I have always been interested in Eastern Woodland cultures and the Hopewell cultures of southern Ohio. As I researched these cultures, I became intrigued by their farming, and this led to more interest in my Cherokee ancestors (of southeastern Kentucky) and their farming methods. This all led me to growing heirlooms, mostly eastern Native American and Appalachian heirlooms. I have about twenty different eastern Native American corn types along with a couple of heirloom dent types. I also grow several types of squash, beans, and sunflowers. I try to grow out three different corns each year, being very, very careful about cross-pollination, not only with my own corns, but with neighbors' fields and gardens. My squash are hand pollinated to preserve their purity, and sunflower heads are "bagged." Growing and preserving our edible heritage is a passion I have.

Tony West has contributed the following essay that draws on his expertise.

The History and Development of Corn

TONY WEST

Corn is a unique grain in that it has no close wild relative. In addition, it was entirely developed by humans and cannot exist in the wild without cultivation. Corn's origin, dating back to 5000 B.C., is believed to be in the Mexican plateau or the highlands of Guatemala. The ancestors of corn are not known for certain, but teosinte and gamagrass are both thought to have contributed to corn's development.

Corn was being grown in eastern North America around A.D. 250. By A.D. 1000, corn had become the most important crop in the Late Eastern Woodland cultures that inhabited the Appalachian region. Because of the favorable mild climate and rich soils of the Appalachians, the pre-European-contact Native Americans were able to develop an agricultural lifestyle that was increasingly based on corn, along with beans, squash, and sunflower.

Corn, more than any other food plant, became the most important base of life for the Native Americans. The Cherokees, who lived in the central and southern Appalachian mountain region after the days of the Eastern Woodland cultures, grew two basic types of corn: flour corn and flint corn. Flour corn was the basis of their day-to-day diet. Flour corn kernels contain almost entirely soft starch, with only a very thin outer seed coat. This type of corn easily grinds into a soft flour and was mostly grown in a solid white variety, though a mixed-colored flour corn was also grown on a lesser scale.

Flint corn, in contrast, has a hard glassy layer entirely surrounding the very small soft inner core and was grown in several colors: yellow, white, red, and a blue-and-white mix. A short-season flint, planted in smaller patches, was used as "green corn," as we do today with sweet corn. The Cherokees were so successful with their agriculture that they usually were able to keep in store up to two years of dry corn. An average Cherokee village would annually grow a thousand acres of corn, interplanted with beans, various types of squash, and sunflowers.

Early European settlers eagerly adopted the corns grown in the area by Native peoples, because these were well adapted to the soil and climatic conditions of the region. Along with the corn crop, European settlers adopted the Native peoples' methods of use, such as making grits, hominy, and corn bread. Settlers were not able to surpass the achievements of the Cherokees in crop production until the various state agricultural experiment stations evolved in the late 1800s, though there were farmers in Ohio, Virginia, Tennessee, and Kentucky working with a new type of corn.

Early explorers collected and traded seed wherever they traveled across the Americas. When an unfamiliar type of corn was found growing in Mexico, seed was obtained and brought back east both by overland trail and through shipping ports in the south. What early seed traders brought with them was a late-flowering type of corn called Gourdseed, which later became known as Southern Dent Corn. It was referred to as a dent corn because the kernel had a pronounced depression, or dent, at the crown of the seed.

Because of their high yield, these new dent corn varieties were traded and grown all through the Appalachian region. By the middle part of the 1800s, crosses between these late-season southern dents and the more familiar local "northern" flints occurred. Successful farmers selectively bred these crosses into new open-pollinated dent varieties with higher yields than either parent stock, with shorter growing seasons than the original Gourdseed had. In time these became what we know today as Corn Belt dents, and they account for about 50 percent of today's world corn production and nearly 90 percent of North America's corn production.

The origin of sweet corn is uncertain. Sweet corn started as a genetic mutation transforming the kernel's starch to a sugar. The low starch levels make the kernel wrinkled rather than plump once the kernel has dried. Archaeological and biological research in recent years points to not one common origin of sweet corn but at least five independent origins.

The earliest records of sweet corn in North America refer to an Iroquois selection named Papoon in the 1770s. Seed listings in the 1820s show a variety simply called Sweet Corn, followed by Darling's Early in 1884, Stowell's Evergreen in the 1850s, and Crosby in 1867. Country Gentlemen was introduced in the 1890s, soon followed by Golden Bantam and Luther Hill, both in 1902. Sweet corns did not attain great popularity in early Appalachia, however, because of the primary need for flour, flint, and dent corn crops. Dent corn's versatility for making flour, roasting ears, and livestock feed far outweighed the luxury of sweet corn on many Appalachian farms.

Before the mid-1800s, most land was worked by hand. After that, many farmers purchased horse-drawn plows, with John Deere leading the plow manufacturers (having sold more than ten thousand steel plows in 1855 alone). Walking behind a two-horse team, a farmer could plow 1¾ acres a day. The later-developed sulky plow, on which the plowman rode, made work easier and gave him greater control. A sulky plow pulled by four horses could plow more than 2½ acres per day.

Plowing was followed by a spike-tooth harrow, which would break up clods in the field. A week or so before planting, a disk was pulled over the field to kill any new weeds, level the surface, and loosen the surface for planting. One or two discings would leave the field ready to plant. An eight-foot disk pulled by four horses could complete 16 acres a day.

Prior to 1850, corn was planted by hand. In the 1700s, corn was routinely planted in raised "hills" approximately three feet in diameter and spaced four feet apart. After the corn was about twelve inches tall, beans would be planted in these hills. As the corn grew, it supplied support for the bean plants to climb. This method of planting was acquired from Native Americans, who planted pumpkins and squashes between the hills to shade out the competing weeds and utilize the unused space.

Although some seed planters were in use in the early 1800s, mechanical corn planters did not arrive until 1853, when George Brown introduced his invention. Mechanical corn planting greatly increased the amount of area that could be planted in a season. Following hand-planting methods, most early corn planters dropped several corn seeds at a time to reproduce the familiar hill-planting method. The ideal amount of seed dropped per hill was three, while hills were spaced 3½ feet apart.

The hill method allowed farmers to cross-cultivate the field, thereby keeping weeds under control through the season. This method of corn planting was common into the early 1900s. Hand planting in Appalachian regions continued on many farms into the early 1900s because of the cost of planters and the small size of fields being planted, in comparison to the immense fields of Corn Belt farms. By hand planting in rows set 3½ feet apart, one person could complete four acres in a day. By the early 1900s, the fortunate farmer who had a two-row horse-drawn mechanical planter could complete fourteen acres a day.

Candy Roasters

Most people in my home community grew a few pumpkins for jack-o'-lanterns and sometimes for cattle feed, but few grew them for human food. The candy roaster was the winter squash of choice then and now. The candy roaster is widely grown in western North Carolina, northern Georgia, and East Tennessee. It has a sweet flavor and is used for making pies, candy roaster butter, and candy roaster breads and can be used in combination with other foods.

The candy roaster is a very long winter squash, thought to have originated with the Cherokees, and can weigh more than fifty pounds. With the advent of refrigeration, portions are now often frozen in airtight packages for use throughout the year.

Typical ways of preparing and cooking candy roasters are as follows:

C. E. Holder with a thirty-pound candy roaster. *Photo by C. E. Holder (self-portrait)*

1. Cut the candy roaster in half and thoroughly clean out the seeds and surrounding area. Cut in one- to two-inch strips and peel off all of the rind. Cut in small pieces about two inches square and cook until tender. Drain off water and mash up the pieces until there are no lumps. This is good for pies; alternatively, put butter and sugar on it and eat it that way.
2. Another preparation method is to cut the candy roaster in half, then clean out the seeds from the inside. Put the squash in the oven on about 300 degrees and bake until the meat is tender. Then take a large spoon and scrape out the inside. Use for pies or just eat as is.
3. To make candy roaster pie:

   2 cups cooked candy roaster prepared using method 1 or 2 above
   1 ½ cups brown sugar
   ½ teaspoon salt
   1 teaspoon allspice

1 teaspoon nutmeg
1 teaspoon cinnamon
3 eggs
1 small can evaporated milk
2 tablespoons melted butter

Mix sugar, salt, and spices. Add to cooked candy roaster. Add eggs (slightly beaten), milk, and butter. Pour into unbaked pie shell and bake for 10 minutes at 425 degrees. Reduce heat and bake for 35 minutes.

This recipe makes two nine-inch pies.

(The above instructions and recipe are from Jewel Dee McCracken of Bald Creek in Upper Crabtree Community, Haywood County, North Carolina.)

Many people cut raw candy roasters into cubes to be frozen for use later in the year. They can then be prepared in the same ways as described above or as one would cook pumpkins or other winter squashes.

Cucumbers

While not as popular as beans and tomatoes, heirloom cucumbers occupy a special status in the gardens of many people who grow heirloom vegetables. Many of the old-time cucumbers

Fred Beddingfield

are somewhat whitish and very tender. They can be eaten without being peeled, and most have a mild flavor.

The following is a story given to me by Fred Beddingfield, of Zirconia, North Carolina, about his great-grandmother Rosie Queen's seed saving.

Grandma Rosie Queen's White Cucumber Seeds

FRED BEDDINGFIELD

These little white pickler-type cucumber seeds were passed down to my great-grandmother Rosie Queen from her aunt not long after the Civil War. She in turn saved the seeds and passed them down to my grandmother Staton, then to my mother, and finally to me.

Like Grandma Queen, I was raised in the Green River community of Henderson County in western North Carolina. During World War II, everyone had large gardens, and many of the locals had small truck farms. Basically these cucumber seeds were planted in "hills" or small mounds, usually five or six seeds to the "hill" and the vines allowed to run over the ground. Later my father grew commercial green cucumbers in fields and trellised them up like pole beans. I decided to plant my white cucumbers the same way, which makes them easier to harvest and cleaner since they aren't on the ground. The vines really like to climb.

My seeds primarily came from my mother, who would allow a few of the cucumbers to go to "seed." They would get rather large in diameter and turn orange in color. Later they would be cut open, cleaned of pulp, and the seeds air and sun dried on a screen wire. Then they were stored in a cool dry place or freezer for the winter.

My mother sent me some seeds to Athens, Greece, while I was stationed there by the Air Force during the period 1967–1969. The cucumbers grew well in the clay soil in Greece and loved to climb my landlord's fence. The Greeks loved them but laughed at the small size. The Greek cucumbers (Anguri) were 15–18 inches long and big and round. The first time I bought some at the local street market I picked the very slim ones, and they tasted terrible. My landlord said to get the big fat ones, and they were delicious but couldn't compare with Grandma's.

Later I was stationed at Patrick Air Force Base, Florida, and lived in base housing just across US A1A from the ocean. The backyard was mostly sand, but the cucumbers grew well in that soil also while just running on the ground. After retirement I moved back to North Carolina and have had a garden here.

Mother passed away in 1996, so it has been left up to me to save the seeds. One year I could not find my seeds. Fortunately my daughter in Florida had some I had given her, so now I make certain I save lots of seeds. I have also given some to my other daughter in Atlanta. They grew well, but some tame rabbits ate them. My son in California raised some, but his two yellow labs ate them. My father-in-law once grew some of these cucumbers in his small garden and was able to make a "run" of pickles in a five-gallon churn. I prefer to eat them, skin and all, just off the vine in the garden.

I have a small garden, so one short row will provide all the cucumbers we need, plus we always give some to friends. In the winter I put all of the leaves that fall around the house in the garden, usually six to ten inches deep. In the spring I mulch the leaves in the soil with a rototiller. This provides nitrogen and other elements to the soil naturally. When planting the seeds I put a little

10-10-10 fertilizer and later add a small amount of ammonium nitrate/sulfate (34-0-0) to give the plants a growth boost.

When I saw the article in the Berea magazine about Bill Best and his heirloom beans I sent him an e-mail about the cucumbers. Bill was a classmate at Berea and also the Alumni Building director later. He was happy to get the seeds and has been raising them. I want to make certain these cucumber seeds are preserved and passed on for years to come.

PART 2

SEED SAVERS

Brian Best, the author's grandson, shelling beans

Seeds, Family, Community, and Traditions

The little pint jar is one of the last my mother canned. I can't bring myself to eat them, so the jar sits on a shelf in my kitchen as a talisman.

JUDITH MARTIN WOODALL

The can house where my mother kept her freezer and her canned fruits and vegetables has only one small window and one electric light. It is mostly underground, so the temperature is fairly constant, and with the light off, it is quite dark. There one can still find on the shelves many quart jars of beans Mother canned as far back as the 1970s, still looking as though they were canned during the past summer. And until recently her freezer still contained bean seeds going all the way back to the mid-1970s.

Traditions die slowly, and the canning traditions of the Southern Appalachians are still very much in evidence when one visits traditional gardeners. In the past few years, I have visited several people who grow large gardens during the summer and can several hundred quarts of beans to use themselves and to give to grown children and close relatives or to anyone else who stops by

Bill Best. *Photo by Irmgard Best*

and expresses an interest. Others store many bushels of potatoes to carry them through the winter and to give to family members and visitors, as well as to use as seed potatoes the coming spring.

These traditions are also maintained in the homes of mountaineers three to five generations removed from their ancestral homes in the mountains. It seems that having beans growing in the garden, strung up for drying, or being canned on the stove keeps one's mind on family long gone but, at the same time, ever present. A recent exchange of e-mails with Judith Martin Woodall of New York City makes the point:

> Dear Mr. Best.
>
> I was delighted to find your website. Three of my grandparents and two of my great-grandparents moved from Knott County, Kentucky, to northern Wisconsin in 1920. They carried with them a number of bean types, but a fall bean variety were the ones we ate the most of. They called the dry beans "shucky," not shelly beans. My mother and father carried on the tradition. I even grew them in a community garden here in New York. Alas, that land was built on in 2002, and since my mother passed away, the tradition of saving seeds from year to year has fallen by the way. My dad still has quite a lot of seed beans, and I have some from my NYC garden, but I fear they are too old and will not germinate.

In my return e-mail, I encouraged her to try to germinate her seeds and those of her father to keep the family bean going. I also asked that she elaborate on the migration process that she was a part of. Her reply came some time later, with an apology for taking so long to respond and explaining that she also was writing a book, about "the history of riding schools and leisure riding in New York City." Woodall, a manager of the Claremont Riding Academy in New York, wrote the following in her e-mail response, which included pictures of the beans:

> Anyway, it belatedly occurred to me to take a couple of photos of the beans. The little pint jar is one of the last my mother canned. I can't bring myself to eat them, so the jar sits on a shelf in my kitchen as a talisman. The coloring of the seed beans is somewhat distorted by the camera's flash, but I think you'll get the idea.
>
> My paternal grandparents were John David and Sally Ann Martin (née Jones), and my maternal grandmother was Bertha May Ball (née Fugate although my great-uncle spelled it Fuguitt, believing it to be of French origin—and therefore much more romantic). My Granny Ball married Herman Ball, the son of German immigrants to Wisconsin. The Kentucks moved from Knott County, Kentucky, in either late 1920 or 1921 to Florence County, Wisconsin, which is now part of the Nicolet-Chequamegon National Forest. Other families who came from Kentucky at that same time were the Ritchies, Gayhearts, McDaniels, Dobsons, Collins, Wells, and others. I can also claim kinship with Hollidays, Mays, Begleys, Ritchies, Allens, Napiers, and so on and so forth. A stroll through the family cemetery would find names completely familiar to anyone from eastern Kentucky. They traveled by train to Wisconsin and lived communally with other families from Knott County in an abandoned logging camp until they got their house built. They purchased 160 acres of land and farmed

and worked in the woods and lived pretty much as they did in Kentucky.

At the onset of World War II, many children of the original generation moved to Kenosha, Wisconsin, where I grew up, to work in factories like American Brass, and Nash, later American Motors—all gone now. This move is akin to the migration from Kentucky to Detroit at the same period. My father was one of 12 children and all of his brothers save one, the oldest, moved to Kenosha or nearby Racine. My parents didn't move to Kenosha until 1950. By then jobs had pretty much dried up in the north. Dad went ahead of us to get work and roomed with one of the Kentucky families who had moved from the north some years earlier.

My Martins are descended from William Martin (born 1760) and Susannah Meador (born 1764). Both were born in Virginia and died in Floyd County, Kentucky—he is the 1830s and she in the 1860s. They came to Kentucky around 1804 and settled on 150 acres, as the deed says, "on the Right Hand Fork of Beaver Creek." From what I understand they are buried in a Martin Cemetery in Wayland, Kentucky. I haven't been to eastern Kentucky in many years but have flown over going from NYC to Louisville and back. The strip mine sites from the air look like a pox on the landscape—deeply disturbing.

Later the same day, Martin sent a follow-up e-mail with this additional information:

I have a sister living in Utah who has not had much success trying to grow the family beans, but I think if they'll grow in a vacant lot in Manhattan, they'll grow anywhere. Of course, our weather is closer to that of Appalachia than is Utah's, and I had the advantage of some excellent organic fertilizer from the stable.

Northern Wisconsin actually has a number of settlements of people from eastern Kentucky, all having migrated there after World War I. Crandon, Wisconsin, has a "Hillbilly Days" every summer. The natives referred to us as "Kentucks," and they didn't always mean it in a friendly way. There's even a lake not far from my dad's named Kentuck Lake. It's rumored that among other things northern Wisconsin was an excellent refuge for moonshiners on the lam from Kentucky. One of my brothers is the owner of the family farm, but he doesn't work the land except to keep a garden (sometimes) and hunt, and pick wild strawberries, raspberries, blueberries and blackberries in the summer. My father lives on a four-acre section, in a house of logs and stone that he and my mother built with help from my brothers.

The nearest town for shopping is Iron River, Michigan, an old mining town that once possessed the deepest iron mines in North America. They didn't tunnel under a mountain; they went straight down into the earth. Those mines have been closed for at least 40 years or more, but the remnants are everywhere and subsidence is a terrible problem.

The Martins' World War II–era migration northward is a common theme among people with roots in Southern Appalachia. Similarly, Harriette Arnow, in her prize-winning book *The Dollmaker,* has her main character, Gertie Nevels, constantly thinking about her home gardens during World War II while trying to adjust to industrial Detroit, where her family had migrated for her husband, Clovis, to take a wartime job. Having to shift to industrial living patterns in a housing project, Gertie feels her world collapsing as she is forced to buy bad food with wartime ration cards and has no garden to sustain her family and her own sanity. Each day she daydreams about what she would be doing back home in Kentucky and even dreams about which bean varieties would be ready to pick at a particular time and when the fattened hogs would be ready to butcher. Time, for her, revolves around the seasons, her gardening and farming, and family and neighbors nearby; time measured by a cuckoo clock is oppressive.

Being able to grow familiar varieties and have a taste of home is a stabilizing force for displaced families such as the Martins. When those family seeds are lost, individuals may go to great lengths to restore them. I get several letters and e-mails daily inquiring about particular beans or tomatoes. Many of those letter writers are anxious to find beans or tomatoes that had been in their families for generations but were lost for one reason or another. Almost always there is some sad family dynamic involved when seeds have been lost.

One of the saddest stories has to do with conflicts within a family in which an elder has died and one of the descendants has thrown away that person's seed collection. Sometimes it has to do with jealousy in a family: the seeds are destroyed to keep them from being divided among other heirs. Sometimes it is done to eradicate bad memories of sibling conflicts or parent/child conflicts. And sometimes it is simply done out of ignorance of the value of seeds.

But most often seeds are seen as a unifying and even uplifting force within a family, a way of keeping in touch with dispersed family members and honoring one's lineage. So it is with great pleasure that I have often been the recipient of a family's treasured seeds whose keepers want them to be shared with others. I have received beans from Washington State when their origin was in West Virginia. I have received beans from Minnesota when the origin was Clinton, Tennessee. I have received beans from Idaho and Ohio and many from Indiana when the origin was Kentucky, and I have received beans from Oregon when the origin was West Virginia. I have met very few people not actively wanting to share their seeds.

I have also received many stories to go with the seeds. Some stories are fresh in people's minds, while others have been passed down through generations. There are stories about young children who grew tired of planting beans and dug a hole next to a stump to hide them, only to later find that they had germinated and left evidence of childish neglect of duty. There are stories of people who refused to share their seeds, only to have them stolen by a neighbor who grew them and distributed them widely. There is the story of the farmer who hauled his beans many miles to Knoxville, Tennessee, to sell at a farmers' market. He would lie down in the wagon and sleep all the way home, because the horse knew the way and also knew that it would be well fed upon its return.

Heirloom seeds and the people who have grown, maintained, distributed, and loved them have been a vital part of my life since I was a tiny child. These seeds have been a part of my extended family and a part of my history. I appreciate every day the roles they have played and continue to play in my life. And I appreciate very much the work and dedication of my fellow seed savers, some of whose stories will conclude this book.

Keepers and Distributors of the Seeds

In addition to the thousands of individuals in the Southern Appalachians who dutifully plant, tend, harvest, eat, preserve, and save seeds of the heirloom and heritage varieties of vegetables, grains, and fruits of the region, there are many who have taken upon themselves the additional responsibility of saving our edible plant genetic heritage for future generations. The following pages honor a few who have passed on recently, as well as some of those still at work, many who have been seed savers for most of their lives.

My mother collected seeds all her life, as her mother and grandmothers had done before her. When in her early fifties she finally got a freezer, seed saving became much easier, and saving seeds from each variety each year was no longer necessary. She freely and rapidly began to acquire even more varieties from extended family members who lived in other western North Carolina counties. Although nearly everyone by that time received gardening catalogues through the mail, she and most of our extended family and neighbors continued to save seeds because quality was "still job one" when it came to food preparation and preservation.

As I entered adulthood, I encountered many other dedicated keepers and distributors of heirloom and heritage seeds. I met many of them at the new Western North Carolina Farmers Market in Asheville, where the beans of my youth were a big hit among local people and travelers to the mountains as well. At the farmers' market, I found myself always headed to one vendor in particular. That vendor was Clive Whitt.

Clive Whitt—the Dean of Beans

On my first visit to the farmers' market, I walked up and down the aisles to check out what most of the vendors were selling. Several were selling heirloom beans, primarily different varieties of greasy beans, but not all vendors were completely familiar with the fruits and vegetables they were selling; they could not really give good answers to my questions about their beans and tomatoes. Clive Whitt stood out from the crowd.

Whitt was retired from a distinguished educational career in Madison County, North Carolina, but he had no intention of sitting around twiddling his thumbs during retirement. He opened one of the first stalls at the new farmers' market and was quite familiar with the products he sold. When I asked him questions, he did not send me somewhere else to find an answer. He spoke very authoritatively, especially about beans.

When I told him I was raised in adjoining Haywood County and had an experience of many years with heirloom beans, he invited me to sort through the beans he had available and buy those that were close to being ready for seed, what we both called "shelly beans." I would spend a half hour or so picking out the yellow-hulled beans from several crates to save for seed. He would then weigh them up; I would buy them and be on my way with still another greasy bean variety to add to my growing collection.

Each time I visited the market in subsequent years I would visit his stand; often I would find still another variety I had not previously seen. I would go through the same process of picking out the most mature bean pods, talk with him for a while about where the bean was grown and by whom, and then be on my way once again to dry out the pods, shell out the seeds, put them in airtight containers, document where they came from, and stick them into one of my freezers.

He occasionally would mention varieties that he rarely had for sale and tell me where in surrounding counties those varieties might be found. One variety he mentioned was the Lazy Wife Greasy from his home county of Madison. The following year, at the March 1998 funeral of my uncle John, I was introduced by his daughter, Linda, to her first cousin on her mother's side, Mary Metcalf. Mary said that she knew where to find the Lazy Wife Greasy and would get some for me. She soon sent me about three hundred seeds through the mail, followed not long afterward by another packet of a smaller version of the Lazy Wife Greasy. The two beans are clearly among the best around and among the best sellers my wife and I send out each winter and spring throughout the United States and Canada.

Clive also had an influence outside of North Carolina. In addition to selling his seeds to me, he also sold them in nearby Tennessee to growers who would then sell him their beans for resale at his stand. Writing in the July 15, 1995, issue of the *Greenville Sun* (Tennessee), in an article about Clive, Bob Hurley made these statements:

> For 43 years he was a well-known and respected educator, mostly in Madison County. The last 15 years of his career saw him as principal of the high school in Marshall. But he got tired of "being daddy" to 800 kids, so he signed out of school and on with Social Security.
>
> "Since I quit the school business, I've been working seven days a week, selling greasy beans and whatever else people will buy," he said. They will buy plenty from Professor Whitt, and he sells it to them for around 10 hours a day. That's 70 hours a week for a man past 82.

In 2009 I was visiting growers in several states to again interview them for this book and was focusing mostly on people in their nineties. (I would focus later on those in their eighties and even later on those in their seventies.) On a visit to North Carolina, I again visited Clive Whitt's stand at the Western Carolina Farmers Market, only to find that he had died two weeks earlier at age ninety-six, more than twenty years after I had first met him.

Upon reflecting on his death, I realized that anyone visiting his stand at the market who took the time to talk with him would come away more educated. It was his nature to share his knowledge with others, a trait that set him apart as an educator and as a vegetable vendor. The current popularity of greasy beans throughout the United States is a tribute to Clive Whitt. He certainly knew about beans, more so than most anyone else.

Eugene M. Parsons

More than ten years ago, Harold Jerrell, the extension agent of Lee County, Virginia, took me to the home of Eugene Parsons in Pennington Gap, Virginia. This was one of my favorite trips while I was getting to know the old-time seed savers and trading seeds with them. Mr. Parsons knew the story of each of his beans and delighted in telling those stories to anyone who would listen.

Unfortunately, in 2007, when I started revisiting my sources for this book, he was in bad health and unable to meet with me. I did talk with his wife, Lillian, on the phone, however, and his son, Herman, and I corresponded by letter. (I had known Herman when he was a student at Berea College.) What follows is a letter Lillian wrote to me on December 17, 2007, giving me some of the details of Eugene's very interesting life. She also sent me some photographs of his beans and tomatoes.

> Eugene M. Parsons was born December 2, 1920, in Lee County, Virginia, in the small community of Puckett's Creek. He married Lillian Cooper on September 6, 1947. He is a veteran of World War II, serving from October 1942 through December 1945. He is the father of five children, grandfather of nine, and great-grandfather of eight so far. He is the third of eight children of Fred L. and Bettie (Howard) Parsons.
>
> Eugene has been a gardener since he was a small boy, helping his mother and siblings clear new ground each spring to expand garden space. He has many family heirloom beans and tomato seeds that he saves from year to year. Some he doesn't even know the names of so we name them ourselves. Such as "Parson's Delights," "Jack and the Beanstalk," which gets its name from climbing a tree so far Eugene had to use a ladder to pick them. One more, "Gigglers," for his brother-in-law who loved beans and laughed a lot.
>
> Eugene is unable to garden any more due to a broken hip and escalating Alzheimer's. He was always an outdoors type of person. He loved gardening, hunting, and fishing. I, his wife of 60 years, Lillian, am writing this for him. Thank you!!

Mr. Parsons died in 2008 at the age of eighty-eight. His love of gardening was very evident when I last visited him, and his seeds will continue to be spread.

Donald Fox

Donald Fox, like Clive Whitt, is from Madison County, North Carolina. A classmate of mine at Berea College, Kentucky, in the mid- to late 1950s, he is also interested in gardening and saving the old seeds. Through a mutual friend, Ben Culbertson, several years ago Don sent me some of his family's greasy beans. We later discussed how this greasy bean came to be.

At the time of the American Revolution, one of Don's ancestors arrived on the coast of North Carolina and ended up fighting on the losing side. At the end of the war he was given a choice of going back to Scotland where he had come from or going into the wilds of western North Carolina. He chose the latter and ended up marrying an Indian woman, one of whose contributions to the marriage was a white greasy bean that is still in the family some 230 years later. Other aspects of Indian lore remained in the family, and Don says that at one time he knew all the healing herbs in the area by their Indian names, taught to him by his grandmother.

Don is now a distiller of whiskey in Kansas but maintains his bean seeds and other heirloom vegetables and his interest in history. He claims that the whiskey he now makes is the first legal whiskey he has ever made.

Betty Flanagan

About fifteen years ago, Betty Flanagan of Summersville, West Virginia, started being given vegetable seeds by the older residents of her area, those in their eighties and nineties. They sort of chose her to be the keeper of their beans because they "weren't going to do anything else with them."

Over a period of fifteen years, she has collected about thirty-five varieties and started marketing them on the Market Bulletin of the West Virginia Department of Agriculture. Many people not from West Virginia receive the bulletin online, so she gets orders from throughout the country. Her orders increase each year.

Now in their mid-sixties, she and her husband, Roger, raise vegetables for seed on about five and a half acres of land. As Betty says, she "loves collecting seeds, which leads to meeting interesting people." She is known for having many bean varieties, including the Logan Giant and the Fat Man beans, which are well known in West Virginia and increasingly in many other states as well.

Harold Jerrell

Harold Jerrell of Lee County, Virginia, retired in June 2010 as an extension agent after twenty-seven years of service. During his years of work, he became deeply involved with the fruits and vegetables of his home county, not letting his education take him away from the things he already knew well, which are increasingly coming to be valued by a public hungry for "real food."

Harold started grafting heritage apple trees many years ago and for years has held workshops to teach both young and old the secrets of grafting and budding. (See his essay on heritage apples and grafting in part 1 of this book.) He came to know as many of the old-time seed savers as possible and gave them encouragement to continue. At the same time, he was creating his own collection of heirloom beans and tomatoes.

One of Harold's projects was to visit as many of the seed savers as possible and help them share their seeds and their stories. With Lee County being on the south side of Pine Mountain, which divides Kentucky and Virginia for many miles, he became aware of the similarities of heirloom seeds from both sides of the mountain. He also knew the richness of heirloom food plant varieties in the area of Cumberland Gap, where Tennessee, Kentucky, and Virginia meet. In many projects he collaborated with growers from the state of Tennessee.

In retirement, he plans on continuing his grafting and seed saving and working with nearby owners of garden shops and vegetable gardens. He is a very important person in Virginia, Kentucky, and Tennessee when it comes to holding on to the old vegetable seeds and fruit trees and freely sharing his knowledge with others.

This is Harold's story, in his own words.

Sharing Traditions

HAROLD JERRELL

During the 1960s, I was a teenage boy growing up on a farm. There were no such things as computers or computer games to occupy my mind. However, outside my home was a world to explore in my spare time. Seldom did neighbors post their property [as no trespassing], and there was no fear of harm, so I often left home, with my two dogs tagging along, to explore this vast world around me. I often walked for hours learning secrets about nature and developing a love for plants and animals that would eventually lead to a college degree in forestry.

I helped my parents milk cows by hand, worked in the garden, and like most southwest Virginia farmers, helped with the production of burley tobacco. When I came home from school, one of my responsibilities was to bring the cows in from the fields for milking. One of my fondest memories during the summer months was making a tomato sandwich to eat while I searched the fields for the cows. However, not just any tomato would serve this purpose. My parents grew three heirloom varieties of tomatoes. One was purple in color, one was red, and the other was yellow with red streaks. These tomatoes were large, and one slice would fit all the way across a piece of bread. Add a little mayonnaise, and you had yourself a real sandwich that would tide you over until suppertime.

In late winter to early spring, tobacco beds would be piled high with wood and burned to generate enough heat to sterilize the soil for the eventual sowing of tobacco seed. This was before methyl bromide could be purchased, and the process involved a lot of labor. Weed seeds that were not killed would have to be pulled later by hand. At the end of the tobacco bed, my parents would sow tomato seeds that had been saved from the previous year or given to them by a neighbor. Tomato plants could be purchased at local farm stores, but they normally did not offer heirloom varieties. After graduating from college and getting married, I began to develop an interest in saving heirloom bean and tomato seeds to grow in my own garden.

To assist the shipping industry, plant breeders unfortunately came up with the idea that tomatoes and beans needed a longer shelf life. By having a "tougher" skin, vegetables could then be shipped to faraway places and were less likely to be damaged. Plant breeders were successful, and soon developed tomatoes that were as firm as a rock and had the flavor of cardboard. Some varieties of green beans would meet the same fate. Fortunately, after a few decades, many people decided they want tomatoes and beans that have good flavor.

Today when I visit local farmers' markets, several varieties of beans such as Cornfield, Turkey Craw, Goose Neck, Greasy, Mountain Climber, and others are often available for sale. Greasy

Jason Davis (*left*), writer of "Save a Little Seed," and his parents, Carolyn and James, in their store

beans are always in high demand and will bring the highest prices. Tomato varieties such as Mr. Stripey, German Johnson, Giant Belgium, Brandywine, Mortgage Lifter, and Oxheart are just a few of the tomatoes that are often available.

Jason Davis from Rose Hill, Virginia, along with his father, Jim, owns and operates Davis Nursery. This spring, Jason grew sixty varieties of tomatoes, and the majority are heirloom varieties. People travel from several states to purchase these tomatoes. It is not uncommon for individuals to purchase plants that are too small for sale, but folks are afraid they will not be available when they return.

I hope stories such as these are sending a message to plant breeders that folks want fruits and vegetables with flavor. And if they do not offer these heirloom varieties, customers are willing to search far and wide for those that will.

Julie Narvell Maruskin

In the absence of serious seed saving by the agriculture departments of colleges and universities, leaders in other state, county, and community institutions have moved into the vacuum to get the job done while it is still possible. One such person in Kentucky is Julie Narvell Maruskin of the Clark County Public Library in Winchester, Kentucky.

Julie Maruskin's parents were from Harlan County, Kentucky. They moved to Paris, Kentucky, then to Bell County, and then retired to Paris, Kentucky. They gardened all their lives, and Julie developed a love of plants and of gardening during her youth. She started planting her own seeds in 1988 while living in Indiana.

As she moved into what became her field, she says that she "liked books, people, and providing a place for self-education." Currently she is the library director of the Clark County Public Library, overseeing a staff of twenty-six people. Much of her time is spent conducting library programs for gardeners who attend her programs at the Clark County library and other libraries throughout the area.

She conducted her first program in Clark County in 2001 with only three people showing up. By 2006, her programs had become so popular that in that year she did sessions at thirty-two libraries, arboretums, and folk arts centers, with over 2,000 people in attendance. In 2009, she had 1,400 people in sixteen sessions. Her programs include seed-starting workshops and seed-saving workshops to assist people in relearning skills that once were taken for granted.

Sensing that many of the heirloom fruits and vegetables she had heard about were in danger of becoming extinct, she has made a special effort to rescue some of the tomatoes she has learned about in her work. These tomatoes include the Butler Skinner Tomato from Winchester, Kentucky, Depp's Pink Firefly from

Julie Narvell Maruskin

Glasgow, Kentucky, and Rebecca Sebastain's Bull Bag Tomato from Paris, Kentucky. She has made seeds of these tomatoes widely available through her library programs and other speaking engagements. As Julie learns more about Kentucky heirloom beans and tomatoes, she tries to document the history of each and often passes out flyers containing that information to those attending her sessions.

As Julie works with those who attend her sessions, she finds people interested in saving money, starting their own collections of heirloom fruits and vegetables, and having more access to food in a purer and less processed state and with fewer chemical alterations. She also finds them becoming increasingly interested in the historical and cultural aspects of their own heirloom foods and also willing to try out native seeds in dooryard gardens and other types of gardens of Greek, Indian, and other nationalities.

Julie's husband, John Maruskin, who also works at the library, often travels with her and makes presentations on planting by the signs. John, a native of western Pennsylvania, is a lifelong gardener who uses signs for much of his planting. He has developed handouts that he gives to participants in their programs; these handouts detail which plants to plant in certain phases of the moon, that is, what to do during the waxing of the moon and what to do in the waning of the moon. For example, aboveground plants should be planted when the moon is waxing (becoming fuller), while belowground plants should be planted when the moon is waning (moving toward the new moon.)

According to research done by Gerald Milnes of the Augusta Heritage Center in Elkins, West Virginia, German settlers in the Southern Appalachians brought a highly sophisticated system of using astrological signs for growing crops. Long a part of German folklore, planting by the signs is still very much a part of the gardening and farming traditions of many traditional gardeners in the Southern Appalachians.

David W. Bradshaw

While working with heirloom fruits and vegetables over the years and while contacting and visiting others who are involved in saving the fruits, vegetables, grains, and other foods of our past, I have noticed that experiences as a very young child are often what set in motion a lifelong desire to save seeds and otherwise preserve edible plants. This is certainly true of David W. Bradshaw, professor emeritus of horticulture at Clemson University. After a long telephone conversation with him in 2010 about his work with saving seeds, I asked him to write a first-person account of his life's work with edible plants. This is his story, told in very eloquent terms.

My Passion for Collecting Heirloom Seeds

DAVID W. BRADSHAW

I have been fascinated with growing plants for as long as I can remember anything in my earliest childhood memories. I grew up in the rural Tidewater area of Virginia. As the third of six siblings of a sharecropping family, I always had a list of chores. Even before I started the first grade of school I milked four or five cows each morning and night. I also enjoyed collecting the eggs from the many hens that free-ranged around the farm. But my favorite pastime was helping Mama and Daddy in the extensive vegetable garden we maintained throughout the year. We grew most of what we ate at the table, including meats, fruits, and vegetables. In addition, Daddy was an avid hunter and fisherman, and he supplemented our diets with wild game and fish he harvested from the woodlands and rivers surrounding the farm we worked.

When I was only five years old and too small to work in the fields, I remember collecting small field corn seedlings that had volunteered in the cotton field. The field workers were removing these unwanted plants from the cotton. I carefully replanted the corn seedlings in the middles between the rows of small peanuts. Daddy was amused at his little farmer growing his own field of corn. He never expected the plants to live. But they did! When Daddy came to cultivate the peanut field, I was digging up my struggling corn plants once more to plant them among the soybean rows. Daddy was certain this time the little corn plants would not survive. I made many trips from the artesian spring with my sand bucket to water my corn. Again they survived and continued to grow, but in the wrong place again. So Daddy took time out from his work to plow up a space in the mulch and humus where we had fed beef cattle all winter. He constructed a fence to protect my plants from the farm animals.

"Now, son, plant your garden here," he encouraged. "That rich soil should grow some mighty fine corn." I was delighted to have a garden all my own. My corn crop grew tall and strong as I weeded it daily and hauled many buckets of water. I still recall how important I felt when Mama and Daddy invited the whole extended family to have an outdoor corn-boiling picnic around the big black wash pot over an open fire.

Soon after our family had shared in the green corn picnic, my maternal grandfather pulled me up onto his lap for a man-to-man talk about family treasures. He told me about the Great Depression and how difficult it had been to feed his large family.

"I had to sell my gold pocket watch to put shoes on the kids, so I don't have that to pass down to you. But I do have this treasure for you."

He opened my hand and placed some snow-white butter bean seeds in my palm. "These are Willow Leaf butter beans that have been grown in our family for many generations. If you grow these butter beans, you will never go hungry. They carried us through the Great Depression. I think you have a green thumb, and you can grow anything. I'm going to depend on you to keep this family treasure growing."

Now, sixty years later I still grow Willow Leaf butter beans. I guess that was my first experience growing heirloom vegetables. I was fortunate to have parents and grandparents who encouraged me by talking to me as an adult. They patiently answered my many questions and challenged me to think. They gave me responsibilities early and expected me to fulfill them. I grew up working hard to grow food for the family, and I always enjoyed most the vegetables I grew in my own garden.

Both my parents were young children during the Great Depression of the early 1900s. They grew up knowing the value of a dollar. They taught me the importance of saving a penny. Every little runt pig born on the farm I rescued and fed with a bottle and, later, scraps from the table and with plenty of buttermilk. When they were ready for market, the pigs were sold, and I was allowed to put the money in my own bank account.

"I am going to buy me a farm someday," I announced proudly.

When I was six years old, Daddy asked me if I wanted to buy into the farming partnership by loaning him $150.00 from my savings account. He used the money as a down payment on a 1951 Dodge pickup truck. That autumn when the crops were sold, Daddy replaced the loan in my savings. Eventually, that early savings account paid for my first semester at North Carolina State University.

I never lost my interest in growing vegetables, and I continue to maintain a large vegetable garden today. Heirlooms have always had a prominent place among my favorite things to grow. An Egyptian walking onion can be traced back almost 150 years in my wife's family. We have grown them for 42 years and shared them with everyone.

While conducting sustainable vegetable production research in the South Carolina Botanical Garden in Clemson, I decided I should plant some of my heirloom vegetable varieties nearby to attract visitors into the area. Almost overnight the heirloom varieties became a hit and were in great demand on my list of speaker's topics. Invariably, my enthusiasm for heirloom vegetables became infectious, and I was offered samples from the gardens of people in the audience. My collection grew rapidly and soon exceeded the available space. After seven years of replicated vegetable research, I planted the whole two and a half acres as an heirloom vegetable demonstration area. I practiced what has become known as memory banking by keeping the family heritage and history with each variety as they were donated and placed in the demonstration garden.

Once more, interest in my heirloom vegetable collection grew as thousands of visitors each year came to visit the area. Soon requests for heirloom seeds were pouring in, but I had no facilities to process and store the seeds we were growing. I visited with Mike Watkins with the South Carolina Foundation Seed Association on the Clemson University campus. Together we forged a partnership so I could produce, harvest, and store many varieties of heirloom vegetables and make them available through the SC Foundation Seed Association outlet.

Now I am retired from Clemson University, but I still grow numerous heirloom vegetables in my home garden. With the passing of time, the SC Foundation Seed Association has ceased operation, and Mike Watkins has also retired. Fortunately, many of the heirloom vegetable varieties from the Bradshaw collection are available through Heavenly Seeds, LLC, located in Anderson, South Carolina.

James R. Veteto

James Veteto, currently assistant professor of anthropology and director of the Southern Seed Legacy at the University of North Texas, has a long history of working with heirloom fruits and vegetables. He is widely published in the area and was a coeditor with Gary Paul Nabhan and Regina Fitzsimmons of the recent *Place-Based Foods of Appalachia: From Rarity to Community Restoration and Market Recovery*, published by the Renewing America's Food Traditions alliance. He still spends his summers growing rare and endangered heirloom vegetables. What follows is his story of his seed-saving adventures, told in his own words.

Appalachian Seed-Saving Experiences

JAMES R. VETETO

I have been a seed saver since 1996. During that year I became aware of the importance of seed saving through the Southern Seed Legacy Project at the University of Georgia run by Robert Rhoades and Virginia Nazarea. It was then that I began to grow out old-time heirloom Southern vegetable varieties and pass them around the South, with as many of the cultural histories attached to them as I could gather. In 1999, I moved up to western North Carolina and ran heirloom vegetable gardens at several locations, including Mountain Gardens and the Arthur Morgan School in Celo, North Carolina. In 2003, I enrolled in the master's program in Appalachian studies and sustainable development at Appalachian State University in Boone, North Carolina, and produced a master's thesis entitled "The History and Survival of Crop Biodiversity and Strategies for Conservation in the Southern Appalachian Mountains of Western North Carolina." During that study I interviewed 27 growers and documented 128 folk crop varieties, many of which I also procured seed samples of. In 2005, I came back to the University of Georgia to pursue a degree in environmental anthropology, which I finished in the spring of 2010. My dissertation is entitled "Seeds of Persistence: Agrobiodiversity, Culture, and Conservation in the American Mountain South." During this project I compiled master biodiversity inventories for the southern/central Appalachian and Ozark mountains and interviewed sixty heirloom growers. Currently my master list of folk variety taxa for central/southern Appalachia contains 1,756 varieties, making Appalachia the region with the highest known agrobiodiversity levels in the United States, Canada, and northern Mexico. In my dissertation I also investigated farmer decision making and how it is related to the continued persistence of heirloom cultivars. My results indicate that heirloom varieties in the Mountain South are still being grown for reasons that can be interpreted as cultural in nature, such as their continued usage in distinct regional culinary traditions and their importance to family history and cultural memory. Mountain South home gardeners can be seen as giving a performance of cultural identity in their agricultural acts, which includes everyday resistances against much of what constitutes modernity in the present day. I also studied and/or collaborated with a number of conservation programs operating in the Mountain South in order to strengthen conservation efforts; these programs include the Center for Cherokee Plants, Conserving Arkansas' Agricultural Heritage, Renewing America's Food Traditions alliance, the Ozark Seed Bank, the Southern Seed Legacy, the Cherokee Nation Seed Bank, and Sustainable Mountain Agriculture Center, Inc. Recently I was named director of the Southern Seed Legacy project, which for me completes a fourteen-year historical loop whereby I now have the honor of leading the organization from which I got my start in seed saving.

I have encountered many interesting characters in my seed-saving adventures and have run across a slew of good stories. An oft-repeated Appalachian tale is that of seeds that originated by hunters killing an animal, finding the undigested seed when they were cleaning the innards of the animal, and subsequently planting it out in their garden and maintaining it for generations in their family. The famous Turkey Craw and Goose Beans are such varieties, but I have also collected a Squirrel Bean and a Wild Goose Corn in western North Carolina that carry variants of the hunter origin tale.

The Eastern Band of Cherokee Indians, most of whom live on the Qualla Boundary reservation in western North Carolina, deserve credit as the originators of much of the indigenous heirloom variety diversity in Southern Appalachia. This is particularly true when speaking of bean, mountain butter bean (*Phaseolus coccineus*), and squash varieties. While conducting research in collaboration with the Center for Cherokee Plants, I have been particularly impressed with the diversity of Cherokee tender October beans and Cherokee butter beans. Roy Lambert, one of the expert seed savers of the tribe, has collected over fifty different bean varieties from the Cherokee Indian Fair Agricultural Exhibit over the years, most of which are varieties of October and butter beans with an amazing array of sizes, colors, and shapes among them. Many Cherokee bean varieties are used to make the famous Appalachian dried and strung beans called leather britches, and this old preservation technology more than likely originated with the Cherokees. Fred Lunsford, an Eastern Cherokee elder and Baptist preacher, told me a hilarious story about leather britches that his wife had preserved and prepared from a Yellow Hull Cornfield Bean that the Lunsfords had originally acquired from Medford Brown (Mrs. Lunsford's grandfather) in Clay County, North Carolina. In 1995, unfortunately Fred Lunsford had a heart attack and was asked by a dietician at the hospital to record the foods he was eating at home. The dietician was from the North. Day after day, as Mr. Lunsford was recording the foods he ate, prominent on the list were leather britches. This greatly puzzled the dietician from the North, who couldn't figure why in the world Fred would be eating his leather britches. Well, she tried to investigate by asking the nurses on duty, but Mr. Lunsford was on to her confusion and told them not to tell her what leather britches were. Finally the dietician asked a Native lady who cooked in the hospital if she knew what leather britches were, and the cook replied with, "Boy, I reckon I do. I'd like to have me some right now." A couple of weeks later Mr. Lunsford was sitting in the rehab clinic when the dietician came in and saw him and said, "Good morning, leather britches." Sometime later, they had a potluck among the rehab patients, and they asked Fred to bring leather britches to share. His wife, Gladys, soaked them overnight in water, washed them in the morning, and cooked them slow with a ham hock. At the potluck they didn't last long, because the Appalachian people who were there knew that properly prepared leather britches have a taste like no other.

There are many varieties around the region that I have documented that are particular to certain families, making them extremely rare, or are adapted to particular mountain microclimates. Jack Banner, a gardener from Watauga County, North Carolina, who was ninety years old when I interviewed him in 2005, had been maintaining a curious variety of potato called New York Pide that has been grown in his family since at least 1892. The Banners were among the original settlers of Sugar Mountain (now a ski resort) near Banner Elk, North Carolina, in Avery County. It is a small, oblong, white potato with a unique taste that is not very productive. Because it didn't produce well, none of the Banners' neighbors in the Sugar Mountain community grew it, but the Banner family liked its unique taste, so they continued to maintain it. Furthermore, when Mr. Banner tried to grow the potato at lower elevations near Marion, North Carolina, when he moved there to take a job in a hospital, he found that it would not grow well there (and I encountered the same phenomenon when I tried to grow in Athens, Georgia; it produced all vines and no tubers). So it is very likely that Mr. Banner has been maintaining a one-of-a-kind heirloom potato variety that is adapted to the higher elevations of NW western North Carolina.

Another rare heirloom variety from western North Carolina is the Roughbark Candy Roaster squash. I collected seeds of this variety from the Bradford family of Bald Mountain, North Carolina. Ernie Bradford told me that he thinks it is a more primitive type of candy roaster—a special type of Appalachian heirloom squash that is thought to have originated with the Cherokee. More than forty different candy roaster types have been identified by undergraduate researcher Jeffrey McConnaughey of Warren Wilson College. The Roughbark Candy Roaster has a thick skin with ridges on it that makes it look "rough" and improves its storage qualities over the winter. Ernie Bradford thinks it has a much richer and sweeter flavor than the more widespread "slick" (thinner skin with no ridges) roasters. The Bradford family uses Roughbark Candy Roasters in pies and for candy roaster butter, candy roaster bread, and as a wintertime complement to bean dishes. Roughbark Candy Roasters grow very large and can weigh over fifty pounds.

These are but a few of the individuals, stories, and heirloom varieties that I have encountered in my seed-saving adventures across Southern Appalachia. My research results indicate that, mirroring the diversity of bioregional natural flora, southern/central Appalachia is the center of agricultural biodiversity in the United States, Canada, and northern Mexico. To maintain this incredible food crop diversity in the region, it is clear that our society must slow down and appreciate what a great legacy that our elders have left to us. The seed savers that I have interviewed over the past seven years are overwhelmingly elderly (average age: seventy years old) and low-income home gardeners. They are nurturing a biodiverse tradition that has been handed down through many generations. Conservation programs in the region are doing their best to preserve this rich heritage, but these programs struggle with limited resources. It is clear that in order to maintain what is left of this valuable Southern seed legacy, we must start working to create a society which supports and nurtures *in vivo* agricultural lifeways and helps create a cultural climate that will allow younger generations to continue the unbroken, but threatened, ancient Appalachian seed-saving tradition.

Darrel Jones

At age fifty-two, Darrel Jones of Hamilton, Alabama, is one of the youngest of the old-time seed savers. He is also a history buff and storyteller who tries to document his seeds as much as possible.

Given his ancestry, which includes three Cherokee ancestors, it is easy to see why seed saving comes so naturally for him. In addition to his work as a telecommunications engineer, he grows and sells tomato plants and produces heirloom seeds (www.selectedplants.com). The following is his story, in his own words.

Grandma Roberts's Beans—a Voyage of Discovery

DARREL JONES

My name is Darrel Jones, and I am a gardener and sometimes a seed and plant seller. This article is not about me so much as it is about a way of life here in the Deep South. People lived simpler, cleaner, and in many ways harder lives than we know today.

My family history stretches from England, Ireland, and Germany to America with ancestors who came across the ocean and homesteaded in places like Greenwood, South Carolina; Sulphur Springs, North Carolina; and Payne's Cove, Tennessee. Several Cherokee Indians are in my lineage. While I know many of the names of my ancestors, very few stories of their lives survived to modern times. A few of them are worth repeating, such as the tale of Philip Hawkins Roberts, who was such a "bad" man that he was elected sheriff of Grundy County, Tennessee, because folks figured he would be able to deal with the "difficult" people in the late 1800s. Another tale is of John T. Prince, aka William Lawson Bailey, who murdered a man in Fannin County, Georgia, in about 1887 and had to run and hide for the rest of his life. He died in Enigma, Georgia, somehow a fitting resting place.

Several of my ancestors fought in the Civil War. John Wesley Mann of the 47th Alabama Infantry's Company A was captured by the North and held until freed by a prisoner swap and then was paroled at Appomattox Courthouse in 1865. James Layton Bailey was a private in Company B Cherokee Legion, North Georgia.

One of the private family tales involves the Cherokee Removal of 1838, aka the Trail of Tears. About a dozen of the Cherokees stayed in Little River Canyon and lived off the land for several years until they were gradually adopted into the local community. One of the daughters married a white man and had several children. One of them was my great-grandmother.

The removal was widely criticized in the North Alabama region because of the privations it caused. There is quite a bit of history preserved at Valley Head, Alabama, aka Williston to the Cherokee.

A little closer to the present would be a tale of my great-aunt Mae, who dug a trench across the road in front of her house so cars (a newfangled invention at the time) would have to slow down so she could see who was going by. My grandfather H. D. Jones Sr. told the tale as a young boy of finding some really strange tracks on the road in front of his house. He followed them nearly five miles until he caught up with the man on the bicycle. He also told of having a .22-caliber rifle during the Depression era and of going hunting one day with only one shell for his gun. He found a large gray squirrel in a tree and aimed his gun; then he thought about it for a while. If he shot the squirrel, he would have meat for the table, but then he would be out of shells and could not hunt anymore. He finally put the gun down and went whistling on his way because as long as he had that one shell, he was "hunting."

I grew up gardening. Every year my father would plow up an area with the old Farmall tractor, and we would plant beans, corn, tomatoes, cabbages, onions, and other vegetables, which my mother would put in jars or in a deep freeze for winter. We usually planted Kentucky Wonder beans, Marglobe and Rutgers tomatoes, and Golden Bantam Cross corn. These were "improved" varieties which began to supplant the old family heirlooms that were widely grown until the 1940s. Just about every family had a good garden, and it was expected to produce food for the table, not to be some decorative place for flowers. That doesn't mean that flowers weren't grown, just that the vegetables were given an importance most gardeners today don't comprehend.

While I had gardened on a small scale in prior years, my first serious garden was in 1986 on land I had purchased the previous year. I carefully ordered seed packs of hybrid tomatoes, cabbage, broccoli, and such that I could start in trays under lights. My dad brought his Ford Jubilee tractor (replacement for the old Farmall) down and plowed up an area of a couple of acres. I bought corn and beans and other seed from a local store and set about planting my garden. It was maybe fifty feet wide and sixty feet long but became a big source of satisfaction in my life and produced quite a bit of food for the table.

Unfortunately, I was about to meet head-on with one of the modern agricultural world's major problems. I had tons of tomatoes . . . but they tasted lousy. This led to a voyage of discovery and a search for tomatoes that tasted good. The next year, I grew some of the hybrids but also some Brandywine and Cherokee Purple tomatoes. Since then I've grown over five hundred different varieties of tomatoes in every shape, size, and color imaginable. Some look like long pointed peppers, while others look like hearts, boats, or various other unusual forms other than the typical red and round hybrid. I've grown red, pink, yellow, orange, green, purple, bicolor, black, and jet-black tomatoes. Some tasted phenomenal; others were absolute spitters. Obviously I grew more of the good ones the next year!

My grandmother, Orphia Roberts, passed away in 1999. When we cleaned out her deep freeze, we found a package of mixed beans from which she planted her garden each year. I could separate the beans out visually into nine different and distinct beans. Unfortunately I have very little information about these beans except the passed-down knowledge of my mother.

Some of the beans were grown in my grandmother's youth in Payne's Cove, Tennessee. Others were given to her by neighbors and friends over the years. I've grown out these beans several times to maintain seed with interesting results.

Three of the beans are the save variety. They are a tricolor that gives white, brown, and black seed. It does not matter which of the three you plant, you get all three colors in the offspring. One of the beans makes pods an intense deep purple color that is almost black. One of the beans is a deep purple pod about seven inches long that is as good as or better than most of the purple pod beans available. Another bean is shaped like a long kidney bean and makes a long pod with very good flavor. None of the beans are greasy types and none of them are cut-shorts. All of them have unique traits that make them worth growing and preserving for the future.

My grandfather Jones was given some gardening books dated 1910 and titled "The Garden Library." He passed them down to me, and now they sit on the dresser beside my bed so I can look in them at my leisure. Among the real treasures in these books are the old ways of growing and preparing vegetables. Good instructions for building a hotbed are lacking in modern literature, but one of these books contains an excellent description of building a hotbed using fresh manure for heat and window sashes to cover the bed. These types of articles are out of fashion with modern gardening publishers, so the only place to find them is in the knowledge of yesteryear.

Gardeners are the world's greatest sharers. The most common way for beans, peas, and corn to spread outside the commercial seed companies is by gardeners sharing what they grow and like. I've been fortunate to receive several really good varieties from gardeners in North Alabama. This is how I first got seed of Blue Marbut bean, White Whippoorwill cowpea, and Granny Franklin Okra. I would stack these and a few others up against the best commercial varieties available for production, flavor, and usefulness. The reason seeds are shared is often because they are the cream of the crop picked for their value in a region. Many such vegetables are regional favorites that seed companies will not carry because they are regionally adapted and can't, for example, be sold in the Midwest as well as the Southeast. The only way to get these seeds is by talking to other gardeners and seeing what you can find in the way of new varieties. The best gardeners for this are those who have gardened the longest. Some of my best finds came from eighty- to ninety-year-old gardeners.

Deep down inside I know that a way of life is passing away while we sit idle. The major seed companies today are developing vegetables that are good for their profits and produce heavily under adverse conditions but are not necessarily good flavored and certainly don't have the history of the seeds that are passed down from one gardener to another. I choose to do something to preserve this history by collecting and growing the varieties of yesterday and by making seeds available for today's gardeners to see what their ancestors grew and enjoyed.

Brook Elliott

Brook Elliott of Richmond, Kentucky, became involved with heirloom vegetables nearly a quarter of a century ago, while doing research for a colonial kitchen garden. "About 80 percent of eighteenth-century varieties are extinct," he recalls reading. "Talk about agri-shock! I was appalled by that, and by the continuing loss of vegetable and livestock varieties right up until yesterday." He became a proponent of heirlooms and promoted them through his writings, lectures, and workshops, thereby helping to create a climate of acceptability leading to their becoming part of the gardening mainstream.

In 2003 he cofounded the Appalachian Heirloom Seed Conservancy, which he managed until 2007, when the organization

Irmgard Best packaging heirloom beans

folded. During the five years the organization existed, it conducted a seed swap each October; this involved many people from six to eight states. (The seed swap continues to be held the first Saturday of each October at the farm of Bill and Irmgard Best.)

Brook currently runs the historic gardens at Fort Boonesborough State Park in Kentucky. As much as possible, actual eighteenth- and early nineteenth-century varieties are grown there, which he tracks down and cultures, both to show visitors what was grown then and to use in the foodways programs he and his wife, Barbara, handle at the park. In addition to the gardens containing only historic plant varieties, all gardening is done using hand tools and eighteenth-century methods.

Growing heirloom plants, per se, is not what Brook is about. Heirlooms are merely part of the big picture. "We are stewards of the land," he stresses, "not owners of it. Making sustainable, biodiverse choices ensures we fulfill our obligations as caretakers." The following is his story, in his own words.

Collecting Kentucky Heirlooms

BROOK ELLIOTT

I began seriously collecting and growing heirloom vegetables about twenty years ago. There were two impetuses. The first was when John Rice Irwin gifted me with a handful of potato onions. I'd never heard of such a thing, at the time, and was intrigued by the idea of onions that cloned themselves.

The second precipitous event came from my work as a food historian, rather than as a gardener. My wife and I are reenactors, focusing on the trans-Allegheny exploration and settlement period. As part of that, we'd developed some expertise in eighteenth-century cookery and foodways, and even wrote a primer on colonial cooking.

I wanted to put in a colonial kitchen garden, using the actual varieties that would have been grown in the trans-mountain region of Kentucky, Tennessee, and western North Carolina. Much to my shock and surprise I discovered that 80 percent of eighteenth-century vegetable varieties were extinct. What's more, that race toward extinction was continuing at an accelerated rate as hybrids and GMOs [that is, genetically modified organisms] took a larger and larger piece of the pie, and programs like the EC's White List further limited seed availability.

Wanting to do something about it catapulted me into the world of heirlooms, sustainability, and alternative agriculture—to the point where the only things we grow are heirlooms and other open-pollinated varieties. I won't put a hybrid in the ground, as much for political as horticultural reasons.

One problem with collecting heirlooms is that every time you hear about one new to you, you have to have it. Pretty soon you would have to own Rhode Island to have enough growing space. The solution is to specialize.

In my case such specialization took two directions. First, I've always been fascinated with the history and diversity of beans, so it was natural that my collecting efforts were concentrated there. Second, and tied in with my living history hobby, I decided to collect varieties from Appalachia, in general, and Kentucky in particular. This would eventually lead to my founding the Appalachian Heirloom Seed Society, which, sadly, is now defunct.

The so-called Middle Ground [Kentucky and Tennessee] is rich in heirloom vegetables. Isolated until fairly recently, families had a greater tendency to save seed and pass it down, one generation to the next. Couple that with the difficulty of travel, and every hollow and valley was, in effect, a foreign country, with its own favored varieties. In just my first two years of specialized collecting, for instance, I had found more than thirty bean varieties, as well as other vegetable types.

Just as important as the seed, itself, are the stories that go with the seeds. Many of these stories are apocryphal, because it's generally difficult to document anything going back more than three generations. Even so, this oral history often provides insights into people's lives and culture that are unavailable any other way.

Obviously, space doesn't let me highlight all the varieties I've collected. But here are a few that are particularly special to me, either because of the stories behind them, or because of the way the variety came to me.

Leona Dillon Bean—A family heirloom from Kathy Williams, of Viper, Kentucky, this pole bean, which has been in her family for at least sixty years, had been literally missing in action for thirty years. When Kathy was cleaning out her deceased mother's freezer, she found a packet of the seed, which had been lost down in a corner. She tried growing them, successfully, and graciously shared them with me.

Jimmy T. Okra—This okra variety was collected by Lisa Huffman, now of Tennessee, from her grandfather Jimmy T. Morris, who grew it in the Elizabethtown area of Kentucky. It has two pod forms: one is ribbed, the other smooth. One form prefers moist conditions, the other dry, so come what may, a crop is ensured.

Black Mountain Pink Tomato—This tomato variety was collected by Austin Isaacs from a man named Harrison at the London, Kentucky, flea market. Back in 1933, Mr. Harrison's father, returning from a hunting trip, found them growing at an abandoned homestead near Harlan, and they've been grown by the family ever since. The variety was named by Mr. Isaacs for nearby Black Mountain, Kentucky's highest peak.

Whippoorwill Cowpea—Perhaps the oldest documented variety in my collection, they were collected by Melody Rose, of Benton, Kentucky, from Dana Adair Mullins, whose family have been growing them as "stock peas" since settling in the Wingo area about 1820.

Kentucky Flat Tan Cornfield Pumpkin—A sweet, flattish, tan-skinned squash collected by John Coykendall, of Knoxville, Tennessee, and grown by him for the restaurant at Blackberry Farm in the Great Smoky Mountains, this Kentucky heirloom is said, by those who know it, to be the finest culinary pumpkin available anywhere.

Dorothy Montgillion

Eighty-four-year-old Dorothy Montgillion has lived in West Virginia since 1975 and is thought of as an herbalist, an herb woman, and other roles but is really, as she says, "a master gardener who looks for wisdom in the ways of our ancestors." With college majors in agronomy and botany, she worked for the U.S. Department of Agriculture in Beltsville, Maryland, in various areas of plant hormone research and seed testing/germination, as well as in the lab that tested all government food purchases for the military, schools, and other government food programs, for twenty years before moving to West Virginia.

In West Virginia she started a business of growing and using herbs for culinary and medicinal use in addition to making jams and jellies the old-fashioned way, using sugar. Her company, Smoke Camp Crafts, sells her products at numerous gift and craft shops and by mail. She also authored a book, *Modern Uses of Traditional Herbs*, available in shops and by mail. The following is her story, in her own words.

Growing Heirloom Vegetables and Herbs in West Virginia

DOROTHY MONTGILLION

My mother was born in Robbinsville, Graham County, North Carolina. She was one of seven children of Squire Patton and Carrie Killian Harwood. As a child, although I lived in D.C. and suburban Maryland, I spent summers at my grandparents' farm. Grandpa was a successful farmer and mostly self-sufficient, as many were in the days before World War II. He farmed with a pair of mules and the sweat of his brow. Grandma tended the kitchen garden, milked the cow, churned the butter, etc. When we were there in the summer I had chores to do, which seemed like fun for a city girl. I was allowed (note the adult psychology at play) to churn the butter, fill the kerosene lamps and trim the wicks, and help pick the goodies from the garden (after Grandpa showed me the ripe watermelon).

At home we always had a garden. Although my father had a Ph.D., he was always just a farmboy at heart. He grew up near Wilmington, North Carolina. My parents met at college in the middle of the state. My brother and I had to help in the garden although it wasn't as much fun at home as with my grandparents in North Carolina.

When I was a senior in high school I knew that I wanted to attend college but hadn't decided on a major. My biology teacher, Howard Owen, was my mentor. If any student showed interest in biology, he would spend time with them in lab after school doing experiments. There were three other seniors besides me who spent many hours exploring the wonders of nature!

At the University of Maryland I had another mentor, Russell Brown, a West Virginia native and a botany professor. I was twenty-five years getting a degree since I dropped out to get married and raise a family. When I went back to work after a twelve-year hiatus, I was a lab tech at the USDA Beltsville Research Center and was able to finish my degree part time. When I started there, our group was working on plant growth regulators, namely, indoleacetic acid. That was when I began to talk to plants!

The greenhouse work became my favorite. Later I worked in the Seed Lab and learned the intricacies of seed saving. Another world opened up for me! When my husband took disability retirement we escaped to West Virginia. Western North Carolina was too far away from the family in Maryland, so we settled for the mountains of West Virginia and have never regretted the move! Of course we have a large garden, complete with herbs. I became a Master Gardener fifteen years ago, am a founder of the West Virginia Herb Association, and operate a business, Smoke Camp Crafts, making jams, jellies, herbal beverages, and medicinal products.

My memories have been many, beginning with my parents, grandparents, high school biology teacher, college botany professor, and my bosses at USDA, John Mitchell and Bernie Leese. You might say that I just happened to be at the right place at the right time.

So here I am now—growing heirloom vegetables and saving seeds, living the good life in the boondocks and loving it!

Seed saving is an important part of all serious gardening for many reasons. I first learned the importance of it many years ago when working at the USDA Beltsville Research Center in Beltsville, Maryland, as a research assistant (lab technician) in the Seed Lab, in testing germination rates of commercial samples and their purity in order to maintain federal standards. Ever notice the "percent germination" listed on seed packets? While employed there I went to the Federal Seed Bank located in Fort Collins, Colorado, on vacation one year. If you ever have a chance to visit there, by all means, go! This is where samples of all known seed varieties are stored, and is the source for developing new varieties (hybrids).

Originally, people saved their seed from year to year. This was possible because the plants were open-pollinated. With hybrids this is unpredictable, as the possibility of obtaining the same plant again is a gamble, or even impossible, as some hybrids produce sterile seeds or no viable seeds. The biggest drawback with hybrids is that they are often being bred for shipping life, long storage life, or uniformity rather than for edibility. A shining example is the tomatoes offered in the grocery store!

There is a growing network of heirloom seed-saving groups thanks in part to the realization of many gardeners that the beans, tomatoes, peppers, and other varieties that their grandparents grew were better than many present-day offerings.

West Virginia established the first state 4-H camp at Jackson's Mill near Weston. I'm lucky enough to live nearby. The location was the boyhood home of Stonewall Jackson from 1830 to 1842. The garden there is planted and maintained by several local

Master Gardeners and has West Virginia Herb Association members assisting the staff. Only varieties grown during that time period are grown there. It was surprising that some varieties grown then are still being grown today, showing that many of the old varieties have superior qualities. It is of the utmost importance that these tried and proven varieties be maintained as a basis for future varieties.

West Virginia (as part of Appalachia) has many heirlooms still being grown today, heirloom tomatoes being probably the most widely grown. Upscale restaurants are ordering heirloom varieties from their local growers. Many heirloom varieties are available at local farmers' markets. Many websites exist to promote these activities. Culinary schools are actively involved in this effort, as they are realizing the economic and culinary advantages. WVFarm2U.org is a website that connects farmers with potential customers.

Rodger Winn

Rodger Winn lives in Little Mountain, a small town northwest of Columbia, South Carolina, with Karen, his wife of nearly thirty years. He retired from the U.S. Navy submarine force in 2000 after twenty years of service and is currently employed by the local utility company as a licensed reactor operator.

When he is not at work, he spends his time renovating their 120-year-old house and gardening intensively. Rodger is a very active seed saver, continuing a lifelong love of growing plants and saving their seeds. He very actively shares his skills and his knowledge with others interested in growing heirloom varieties and saving seeds. The following is his story, in his own words.

A Jar of Seeds

RODGER WINN

As a small child I was fascinated with watching things grow. While other kids came home from school and played baseball or football, I wanted to help my neighbor in her vegetable garden. I did play Little League sports, but my heart was not into it. Instead I would spend my summers and free time after school roaming the Mennonite farms that surrounded our subdevelopment.

In second grade, as a class science project we studied how plants grew. My teacher had a mason jar that was stuffed with wet paper towels and had butter bean seeds between the glass and the paper towels. For the next few weeks we watched the beans swell, then grow roots and a stem. That year we also grew a sweet potato vine in a jar from a small piece held in place with toothpicks. By the end of the year, the vines sprawled across the large windows that flanked the outside wall of my classroom. I was so fascinated with watching plants grow that I started to run my own seed germination projects. I became interested in gathering seeds from all types of plants and germinating them. I even made a journal where I drew pictures of various seeds and the stages of germination. One year I grew maple tree seeds and sold the plants for a dollar apiece and made nearly $50. But what really started me on growing and fueled my affliction with saving seed was a pumpkin seed I planted in the spring of 1967, and by midsummer had a pumpkin. My dad, Hoyt Winn, finally tired of the vine growing across the lawn and deemed the pumpkin ripe, so I picked it. I was so proud of that pumpkin. Each morning I carried it to the porch and displayed it for everyone to see, and each evening I carried it back inside to my room.

But as fate would have it I stumbled and dropped the pumpkin and it split open. I believe I cried for a week but my mom did make a pie out of it. We lived in Virginia Beach, Virginia, at that time. My dad was in the navy, and each summer we would go to South Carolina to visit my dad's family and to Alabama to visit my mom's family. I really enjoyed going to Alabama. My mom, Jan Winn, grew up on a farm in the community of Helican, which is situated in Winston County in the mountains of northern Alabama. My grandparents ran a dairy and chicken farm and always had a huge garden.

After hearing the story about my pumpkin, my grandmother gave me the best gift ever: a jar of seeds. Most seven-year-old boys would want a toy truck or trading cards; I got a mason jar filled with an assortment of seeds and couldn't have been happier. I had some dried bean pods, an okra pod, some watermelon seeds, a few field pea pods, and some multicolored butter bean seeds. Every day I pulled out those seeds and gazed at the shapes and color patterns. I was fascinated with them. I could not wait till spring to plant my first real garden with my own seeds saved by my grandmother and given to me.

Those seeds have long been lost, but they instilled in me a desire to learn everything I could about gardening and collecting seeds. By the time I was in eighth grade we were living in a subdevelopment in Columbia, South Carolina. I wanted a garden, but we had too many trees and no place to plant vegetables. My aunt Elizabeth knew I wanted a garden and asked me if I wanted to plant a garden next to hers in the country. I was elated. I spent nearly every Friday night with her so I could work my garden on Saturdays. My mom canned a lot of beans, and I sold the excess produce. I had some of the best summers as a young teenager working my garden. My aunt still loves to share stories of how I planted a garden for a couple of summers at her place.

After high school I went to college but was bored of sitting in a classroom, so after a year of college I joined the navy. I became a nuclear operator on a submarine. In 1981 I married a girl I met while I was in college. She grew up next to her grandparents in Little Mountain, South Carolina, where her family grew their own vegetables and more peanuts, sweet potatoes, and watermelons than I thought possible. My wife's grandparents always saved seeds for everything they grew.

While in the navy I had a garden and always brought some of my seeds with me when I went to sea. It reminded me of home and created good conversation. I had been given seeds by my wife's grandparents and family—seeds I still cherish today, such as the Metze white cucumber, Uncle JC's Zelma Zesta bean, and Grandma Wicker's brown and white half-runner bean. I also began to collect seed from people I met while in the navy. I remember a man named Bill Soleau (Swallow). He was a Cajun from Louisiana. We both served together in Kings Bay, Georgia. One Christmas he went home and told his grandmother about me and my old-timey seeds. When he came back he gave me a sandwich bag full of okra seeds and a label that said Louisiana Bull Horn Okra. His grandmother told him to tell me if I was going to grow okra and liked the good old-timey seed then I needed some of the best gumbo okra around and not to thank her for the seeds lest they wouldn't grow but instead plant them and if I liked them, then save some seeds and pass them on to someone else. I kept her wishes and have passed on the Louisiana Bull Horn Okra to probably a hundred people over the years.

In 1996 I bought the house that I live in today. We were home visiting our families after spending a couple of years in Hawaii. During breakfast one Saturday morning, I was looking at the classifieds when I saw an ad for a hundred-year-old farmhouse in Little Mountain. I pointed it out to my wife, Karen, but she didn't know the road. When we last lived in South Carolina, rural roads didn't have names, only numbers. That afternoon I decided to call the realtor to see the house. When I stepped out of my truck I was finally home. That evening when my wife returned we went to the house. She said she knew the house and family and that they were kin to her. I told her I had bought it. I thought she would get mad but instead a smile came across her face. I now had a house greatly in need of repair only four years before I would be retiring from the navy to make it habitable. The garden spot still had multiplying onions and garlic at the entrance, and that first spring larkspur, blue bachelor buttons, and vining petunias came up everywhere. Thousands of daffodils and narcissi outlined all the outbuildings and paths around the home. Not only did I have the house of my dreams, but I also had a garden full of heirloom flowers and a place to plant all the seeds I had collected over the years.

In the spring of 1998 I began to search for butter beans like the ones my grandmother Eula Johnson in Alabama used to grow: the same butter bean seeds she gave me when I was seven. They were colored butter beans, as she called them, and she had several types; some had only one color, while others had multiple colors. It was the multiple-colored ones she liked the best. I can still vividly recall her sitting on the porch on a hot summer evening, with a box fan circulating the air and her shelling butter beans and talking. Every so often she would get quiet and run her hands in the dishpan of freshly shelled beans. I remember asking why she did that, and she replied that she enjoyed looking at all the color patterns. It never occurred to me as a young man that the seeds she cherished would be lost. When she was not able to take care of herself in the early '80s, her seeds died and were thrown out when the house was cleaned and property divided up.

While searching the Internet for multicolored butter beans, I came across a site called Southern Legacy Seeds. It was a seed bank that preserved older varieties of seeds from the South. I joined Southern Legacy Seeds. In the seed savers' memory book was an article on saving butter bean seed by a man named John Coykendall of Knoxville, Tennessee. I contacted him and acquired a pinkeye butter bean seed and a multicolored butter bean just like the ones my grandmother grew. This was the first time I realized that I was not an isolated case: there are other people like me who have a peculiar affection for growing old-timey seeds. I was an "heirloom gardener." I have since come across and joined other heirloom seed groups, such as Seed Savers Exchange in Decorah, Iowa, and the Appalachian Heirloom Seed Conservancy in Berea, Kentucky. Even though the latter is no longer an active group, there is still a dedicated group of growers, collectors, and preservationists who meet each fall for an informal seed swap started by the conservancy and to enjoy a day of fellowship in the Kentucky mountains.

For the past few years I have been invited to Monticello in Virginia for a fall heirloom garden festival where I do a seminar on bean seed saving and facilitate an old-timey seed swap. I brought my son, Christopher, and his wife, Nichole, one year. My daughter-in-law mentioned to me later that she would have never believed that there were other people as crazy as I am about a bean. It just blew her mind to see such a diverse mix of people, from young educated adults to middle-aged men and women, the hippie stereotypes and backwoods hillbillies that all came together to trade seed and swap stories.

I am now retired from the navy and work just twelve miles from my home. I do the same job I did in the navy: operate a nuclear reactor to produce electricity. I enjoy my work; I actually look forward to going back after a long weekend to give my body a chance to rest and to talk about what I am growing. It is odd that I do a job that is in an air-conditioned computer-driven environment of fluorescent lights and control panels. Yet I would rather toil in the dirt in ninety-plus degree heat just for the satisfaction of seeing something grow. I also have a captive audience at work, and just like in my navy career when I brought my seeds to sea with me, I also bring seeds to work, along with fresh fruits and vegetables like yellow-meat watermelons, white cucumbers, black tomatoes, and purple-striped beans. And just like in the navy I have come across colleagues whose parents or grandparents also grow some of the unique vegetables that I have always grown and shared, and they also share seed with me and tell me the stories behind them.

I am still fascinated with germinating seed. I grow about ten thousand plants, mainly heirloom tomatoes but also peppers, eggplants, melons, and some herbs each spring for sale. I have a full acre of land where I grow certified organic seed crops of heirloom varieties for several seed companies and for preservation. My wife and I enjoy going to seed swaps and meeting people just like us who have a passion for growing and preserving the old-timey varieties. I do several talks a year to local garden clubs and community groups about heirloom gardening. Whenever I attend a seed swap or do a seminar, I am always asked the same two questions: Why do I grow heirlooms? And how did I get started? I don't have a good answer for why I grow heirlooms—definitely for flavor but also for the memories I have. I have never grown modern hybrids. I have never had any need to, because I have always had seeds of vegetables that have been passed down to me by my family and shared with me by friends that I have met over the years. As for how I got started, it was as a seven-year-old boy with a pumpkin seed that not only produced a pumpkin but also netted me a jar of seeds from my grandmother and a memory that I will always cherish.

I have a granddaughter, Natalie, and another grandchild on the way. When I have Natalie in the garden with me, I tell her what seed we are planting and where that seed came from. One day I hope all my grandchildren will remember me and my seeds and my love and desire to garden. And when they are school age and ready for their own little spot, I will give each of them a jar of seeds just like my grandmother gave me.

Jack Woodworth

Jack Woodworth

Jack Woodworth and his wife, Andrea, live on an eighty-acre farm in southwestern Virginia where they raise a variety of livestock, concentrating mostly on the Alpine dairy goats used in their Grade A dairy operation. He also grows many heirloom fruits and vegetables in several gardens, using the fruits and vegetables for home consumption and for saving seeds. He sells many of the seeds to a local seed company and also works with and manages a community-supported agriculture (CSA) co-op. He actively supports seed-saving organizations, gives talks to interested groups, and participates in seed swaps throughout a wide area. The following is his story, in his own words.

Seed Saving in Southwestern Virginia

JACK WOODWORTH

My name is John Woodworth, but I've been called Jack since childhood. I'm comfortable with both, and most folks around here know me as Jack. I grew up in Connecticut, joined the army in 1973, and went to Germany in January 1974, returning to Connecticut in 1981 with a German wife and an eight-month-old daughter. She was joined by another daughter in 1983 and a son in 1989. We lived in Bethlehem, Connecticut, for twelve years before moving to Gate City, Virginia. I worked in a zinc rolling mill in Waterbury, Connecticut, as a quality control technician while there. Since moving to Virginia, I have been self-employed as a farmer.

My wife of thirty-four years, Andrea, is an R.N. I grow seed for Southern Exposure Seed Exchange as well as for my own use and for direct sales to anyone interested. I manage a Community-Supported Agriculture (CSA) called Highlands Bioproduce, Inc., which is a co-op of certified organic growers as well as those who grow in an all-natural manner. My wife and I also have a Grade A goat dairy where we presently milk thirty-nine does, making the milk into artisan cheeses, which we market at a couple of farmers' markets and supply a number of "high end" restaurants. We also have a variety of other livestock: Jacob sheep, rabbits, poultry, a pig, and a couple of horses. Add in the farm cats and a bunch of dogs, and we keep pretty busy. I am secretary at my church as well as a Sunday school teacher, and I chair the Scott County Extension Leadership Committee and help out with the 4-H poultry program.

I've been interested in growing plants since I can remember. My maternal grandfather always had the most beautiful flower gardens. I remember collecting seed from his flowering balsam plants as a boy. While I was stationed in Germany, a sergeant with whom I worked got me interested in growing house plants from seed, which kind of got the seed-collecting ball rolling. Unfortunately, all of those plants had to remain when we left Germany. At our first house, in Bethlehem, Connecticut, we had room for a garden, and I got interested in heirlooms and began to collect seed to grow my own, but somewhat haphazardly. What got me really started was when I had saved a bit of seed from a Hogheart Tomato, originally purchased from FEDCO Seed in Maine. I went to order more, but it was not available as they had had a crop failure.

I planted some of what I had and bought Marc Rogers's book on seed saving and began to learn how to properly save and store seeds. I got interested in Appalachian heirlooms after I read an ad for the Appalachian Heirloom Seed Conservancy in *Small Farm Today* magazine. I regularly have a booth at the Exchange Place Spring Fair at which I advertise our CSA, so I contacted Brook Elliott to see if he'd like me to put out fliers to advertise AHSC. I thought it was one of the most sensible ideas I had ever heard of, and hope that it may be revived in some form in the future.

Appalachian Farmers Market Association sponsored a seed swap last December in Bristol, Tennessee, at which time I was the featured speaker. It seems that there is still a growing interest in saving seed, and I'll do what I can to help any interested party to get started. I've spoken to a number of groups about seed saving and will continue to do so when possible. While I own and have read many books on heirlooms, there is no substitute for talking to the people who have grown them for years and saved their seed. I have not found a whole lot of printed information on Appalachian heirlooms specifically; hopefully, this book will fill the void.

There are many of the seeds I have that are not from this area, but I continue to search. There are three on my list that I have named: The first is Charlie Hall Cut-Short pole bean, named for the gentleman who brought the bean to the nursing home where my wife worked to see if his mom remembered what they were called (she didn't). I brought the one bean to Berea with me, but no one there could name it, so I named it for him. The second, Carroll's Calico Lima Bean, is named for a neighbor, Jim Carroll, who got it from a friend who was helping clean out the freezer of the man who originally owned the farm where Eastman Chemical Company now stands in Kingsport, Tennessee. Jim was one of the original "Rocket Boys" on which the book/movie *October Sky* was based. (The character Odell was a compilation of him and another boy.) Jim gave me the seed to bring to Berea for identification. Bill Best told me it was probably from West Virginia and that it was a calico type. John Coykendall offered this seed in this year's Seed Savers Exchange book as J. Carroll, but he got it from me, and I named it Carroll's Calico based on information that I got from Bill Best. The third on my list is Jack's Speckled Giant. This is a mutation that showed up in Logan Giant beans that I got from a friend. It's possible that they are a cross with Turkey Craw, as the seed looks like a Turkey Craw. Although I grew them at the same time, they were isolated from each other by quite a distance, so the cross probably would have been before she gave them to me. She did give me both varieties but swore that they hadn't been planted near each other. I have grown it a few seasons, and it is starting to breed true.

Frank Barnett

Frank Barnett has accumulated an impressive collection of heirloom beans, primarily from eastern Kentucky. Since his retirement in 2007 he has logged thousands of miles in search of heirloom beans and their growers. In Breathitt County, Kentucky, alone he has found 46 varieties and is still looking. Altogether he has found over 150 varieties, many of them of outstanding quality. The following is his story, in his own words.

John Coykendall (*left*) and Frank Barnett (*right*) at Harlan County seed swap

Grandma Barnett's Beans

FRANK BARNETT

A large portion of my childhood was spent in Columbus, Ohio. My father worked for the C & O Railroad at the roundhouse in Columbus, which maintained steam and later diesel engines. He finally got to transfer to the Raceland Car shop, where coal-hauler cars were built, in Greenup County, Kentucky, when we started high school. My sister and I were glad to leave the shotgun house on the south side of Columbus and move back up a hollow in Floyd County. It was a big improvement in our lives.

We didn't have much space for a garden in Columbus, but there were a large number of roadside markets in the countryside. A large population with Appalachian roots meant that white half-runners were the predominant bean sold in the 1950s and early '60s. Canning was a big priority, and usually my mother would can around three hundred quarts. Of course, we also would time our visits to Floyd County, Kentucky, and bring back beans from Grandma Barnett.

In my mother's family there were nine children, and my grandfather worked in the coal mines. They were continually moving, since my grandfather had to move to find work in the mines. They didn't save any seeds, and it was difficult to find a place to rent with nine children. My grandfather finally moved his family to Michigan in 1950.

They moved to the muck farms as tenants. There was quite a population on the farms with others from Appalachia along with migrant Mexican workers. The farms mainly grew onions, potatoes, mint, and sugar beets. The base pay was around thirty-five cents an hour, but the entire family was able to work at least part time. Living and working on the farms provided them with all the potatoes and onions they could eat, something they seldom had in Kentucky. In addition, my grandfather had a full-time factory job.

In my father's family there were eleven children. My grandparents also raised an orphaned boy whose father had been killed in the mines. Later the boy went to work in the same mine as a teenager and was also killed. My grandmother later, as a widow, raised three of her grandchildren. So she raised a total of fifteen children.

Grandma Ella Barnett was the gardener and seed saver. She saved her beans in jars with a mothball or hot peppers, or sometimes she would just tie them up in an old rag. She managed to milk one and sometimes two cows a day, raise her chickens, and raise a hog or two. And she had her dogs, Coaly and Bear. Perhaps all her garden patches totaled an acre or so, but it was hard to determine on some of those hillsides.

I remember Grandma splitting wood with a single-bit axe and firing up that big black pot in the backyard, which backed up to the hillside. It seemed odd to me that lard, wood-ash lye, and water would make soap. Of course, she had used the same pot to render out lard and made the cracklings that she would toss out on an old wooden table, where we would grab them and burn ourselves eating them like candy.

Grandma would can and dry beans for shuck beans. She canned her beans outside in a washtub. I had to call my mother, who is eighty-three years old, for these details. My mother said usually most people, before my time, would have two washtubs going at once to can beans. Wood had to be gathered, which was usually logs, [to build the fire in a pit under the tub. The tub was filled with] water drawn from the dug well, and the tub [was then] placed over a pit [and] supported on rocks. Rags or cardboard were placed between each jar to prevent them from touching and exploding.

Pieces of flat wood had to be placed on the inside bottom of the tub to prevent the bottoms of the jars from touching the tub, overheating, and exploding. And rags were placed on the tops of the jars to keep as much water in the tub as possible. Water also had to be boiled inside the house to pour into the tub to prevent the tub from boiling dry. This boiling usually continued for around four hours. And the result was usually, hopefully, nineteen or twenty quarts of beans.

I remember the wood-burning cookstove Grandma had in the house and the one she also had out in the yard beside the smokehouse. There were nails, probably 20d's, driven along the top of the wall near the ceiling behind the entire wall behind the stove with strings of beans hanging, and you didn't see much wall between those beans.

How anyone could manage to cook or bake with a wood-burning stove is beyond me. But when Grandma was cooking with her inside wood stove trying to dry her beans, we got to go out on the front porch and eat because it was so hot in the house. Green beans, green onions, greasy cornbread, fried potatoes, raw milk, and homemade butter were the norm, along with tomatoes, corn, and greens when in season. Those beans were cooked with some kind of fat pork, and for good measure that big lard can was in the floor and perhaps a scoop or two of pure lard would top them off. And, by the way, I never weighed over 140 pounds.

I've heard lots of people say that the beans dried in this manner were so much better than those dried outside. Seems I remember the kitchen being smoky. And of course with frying pork of some type every morning there might have been some grease in the air. Perhaps that affected the taste of those shuck beans.

Grandma never bought a seed or tomato plant until she was maybe eighty years old. She had always raised plants using her own saved seed. Unfortunately, we let those old bean varieties get away because we took them for granted and thought they would always be around.

Later in life in my work career I discovered some really bad beans. Nearly thirty years ago I accepted a job transfer to the Mid Hudson Valley in New York. The company cafeteria served green beans that I considered to be raw. And I missed my Martha White Cornmeal mix, but I would load up on my trips back home.

After two and a half years I got to transfer to central Indiana. I met an elderly gardener who was originally from Rockcastle County, Kentucky. He raised beans and raspberries to sell. He had given up raising any bean with a string years before because no one would buy them. But he had a lot of stories about Rockcastle County and the times and beans that had been forgotten.

In 1987 I finally got to return to Kentucky. My prime consideration in buying a house was that I wanted space for a large garden with great soil, which I found in Scott County. I started to raise my grandmother's surviving bean variety after she passed away in 1990 at the age of ninety-four. I decided to concentrate on a more vegetarian diet to allow me to deal with the stress of my job.

At first I didn't have much success locating any old-time beans, since my grandmother was the last old-time gardener on the creek. So I started visiting the local farmers' markets with some success. But I made a lot of progress after my retirement in 2007. In 2009 I met a retired schoolteacher in Breathitt County who also had driven trucks all over eastern Kentucky every summer, and he gave me some good advice as to where to travel and where to just stay away from since he had never seen a garden. So, 2010 was a banner year for me.

John Coykendall

John Coykendall was born on April 17, 1943, in Knoxville, Tennessee. He attended the Ringling School of Art from 1962 through 1966 and the School of the Museum of Fine Arts from 1966 through 1970. For several years now he has been the master gardener for the Blackberry Farm resort in Tennessee, where he grows heirloom vegetables for its restaurants. He continues to experiment with heirlooms and is always growing out ones new to him. The following is his story, in his own words.

A Sense of Responsibility

JOHN COYKENDALL

For the past ten years I have been working at Blackberry Farm, a well-known resort located in the foothills of the Smoky Mountains in East Tennessee. Most of my time at Blackberry is spent growing unique heirloom seeds that are sold at the farm and also raising unique produce for the chefs to use in their culinary creations. My greatest passion is searching for heirloom varieties, growing them out, and making the old varieties available for others to grow and pass on.

My earliest memory of being a seed saver goes back to the mid-1950s, when I began saving seeds from the old-time Kentucky field pumpkins that were raised in the cornfields on my grandfather's farm. It seems as though my "seed-saving gene" began expressing itself early in life.

As far back as I can remember, I sought out the company of old-timers in the farming community. I loved their old-time tales and the knowledge that they passed on to me about the old-time farming ways and the old varieties of seeds that had been passed down to them by their fathers or farming neighbors.

If I had to recall a single moment in life when the quest for finding and saving old-time seeds began, it would have to be a spring day in March of 1959 when I crawled through a basement window of an old abandoned railroad station west of Knoxville and found among some old magazines a 1913 seed catalog from the Maule Seed Company of Philadelphia. I still have the old catalog, which is filled with engravings of the old vegetable varieties.

In studying these old illustrations, I was fascinated with so many names and varieties that I had never heard of before. I immediately had so many questions: What happened to these old varieties? How many might still exist? How many had been lost? Where would you begin your search for these old varieties?

In 1959 there was no formal network of seed-saving organizations in the United States, or at least not as they exist today. Possibly the best model for a seed-saving organization, the Seed Savers Exchange, was founded in 1975. That's late in the eleventh hour in terms of collecting and preserving seed, especially when one considers the hundreds or thousands of years that have passed since the dawn of agriculture.

Quilt hanging titled "The Noble Bean," made by Linda Simpson and based on a photo by Linda Simpson. *Photo by Bill Best*

Years would pass before the network of seed savers as we know it came along. Years ago there were seed savers preserving unique varieties, but there was no way of knowing what Bill Best was preserving up in Berea, Kentucky, or what Rodger Winn down in South Carolina had.

People often ask me, "Where do you find heritage seeds?" I don't know of a simple answer. In the past I have collected seed from my farming neighbors or from an individual who has heard that I preserve old varieties. Occasionally I get a package in the mail containing one or more varieties, but in the end I can't claim that I have any special talent for tracking down heritage varieties. A great deal depended on pure luck or being at the right place at the right time.

Whenever I am given seed for an old variety, I always treat it as though I were the only person on earth who had the seed, and

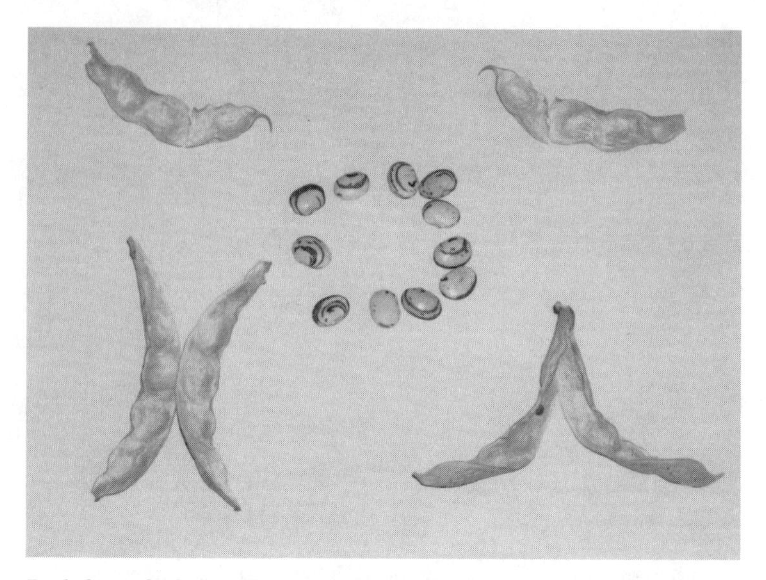

Back from the brink of extinction, the Noble Bean seeds, at one time the only ones in existence

in some cases that might be true. I always feel a great sense of responsibility when seeds are given to me, like having a child. I am totally responsible for the well-being and preservation of these seeds.

Two winters ago when I was paying a visit to Mr. Bill Best's farm in Berea, Kentucky, he gave me one seed from the Noble Bean, an excellent example of a bean that he had rescued from extinction. Getting that bean was like caring for a child. It was started in a pot, nurtured until it was large enough to survive in the garden, and then planted with fencing around it, a pole to grow on, and plenty of observation every day. I don't know how I would have faced Mr. Best if I had failed to successfully grow out the Noble Bean, but as fate would have it, that one plant produced a great abundance of pods and enough dried beans along with Mr. Best's grow-outs to ensure this variety's survival.

Although most of my seed collecting has been done in the southeastern United States, I have also collected quite a few old varieties in Austria where I lived in 1970–1971. In 2009 I spent time in Romania, where much of the country is still farming the way they did over one hundred years ago, and a great diversity of seed is still to be found.

The Important Plant Genetics Work of Kent Poe

Kent Poe of Jefferson, North Carolina, has done significant work with the American chestnut and the Fraser fir. He is also working with a mutant strain of the Kentucky Wonder bean. The American chestnut once dominated the forests of the eastern half of the United States and furnished food for humans and animals alike. Sometimes known as the "Redwood of the East," because of its great size, the American chestnut was largely destroyed by a fungus brought in from Asia in the early part of the twentieth century. Kent now has chestnut trees from his "Mother Tree" growing in four states.

While the Fraser fir is not a food plant, it has a sizeable economic impact in counties where it is grown, and Kent has done significant work with it as well.

Kent's father at one time grew hundreds of acres of pole beans for sale in the South. Kent has continued growing the Kentucky Wonder and maintains the purity of the original seeds, which are still tender, unlike the many commercial bean varieties that have been toughened for machine harvest.

American Chestnut

KENT POE

I started working with American chestnuts in 1976. A natural American chestnut was found in the Little Phoenix area of Ashe County, North Carolina, which had about a half gallon of chestnuts. The tree belonged to Don Witherspoon. My son, Shawn Poe, collected the seeds. An article was written in the local newspaper, the *Skyland Post,* showing Shawn and the seeds he had collected.

My primary effort has been in pushing the generations forward as fast as possible. I came to the conclusion that no genetic resistance to the chestnut blight had been possible, because of the swift death of original American chestnuts. Only by seed production can the chromosomes change to develop disease resistance.

My first seeds were sown in the spring of 1977. From that sowing, about one hundred seeds, I had one seedling that showed great vigor and blight resistance. This tree still lives on my farm in Jefferson, North Carolina; I call it the "Mother Tree."

Some people say it is probably a natural cross with the Oriental chestnut. All other trees from that first sowing had smaller nuts and were much more susceptible to the chestnut blight.

Gwyn Campbell

Gwyn Campbell is, like Kent Poe, a seed saver from Ashe County, North Carolina. Retiring after a thirty-three-year career as a vocational agriculture teacher, he continues to maintain the heirloom beans of his area and also works to locate and graft fruit trees of Ashe and surrounding counties, a labor of love going back many years.

An heirloom half-runner bean from his wife's family is one of the beans he maintains and is one of the best of the type, being very tender at all stages of development and making an excellent shuck bean as well.

Jim Gore

Jim Gore has farmed full time near Peterstown, West Virginia, since his retirement from the public schools, where he spent nine years in the classroom and thirty years as a school principal. In addition to growing many acres of sweet corn, he has experimented with and maintains many varieties of heirloom beans and tomatoes that he shares with others. He sells not only tomatoes for eating but also tomato plants from his greenhouse. His best-selling tomato plant is one that bears his name, Gore's Yellow with Red Center, a potato-leafed yellow German type that mutated from another yellow German variety he happened to be growing. His tomato is much more vigorous than the one from which it originated and has become quite popular. His method of plant breeding is to select outstanding plants from which to save seeds. The following is his story, in his own words.

Saving and Sharing

JIM GORE

I was born on a farm in the Appalachian Mountains thirteen days after D-Day in the midst of World War II. We, like our ancestors, grew almost everything we ate and used. My parents were born in 1893 and 1905 during the agrarian era of American history. Almost everyone lived on a farm and raised large families to help them extract a living from the good earth. I was almost grown before I realized there were some people who didn't live on a farm.

It is a heritage I am proud to own. It is the spirit of self-sustenance which intrigues me. To this day, it is very satisfying when everything on the dinner table is produced at home. I have come to realize that this heritage is quite rich in Appalachia. I like to refer to myself as an Appalachian American.

A friend of mine gave a gardening presentation to a kindergarten class at our local elementary school. He asked them where their food came from. All but one said the supermarket. How sad! The average American is three generations removed from the farm.

Many of those who still raise their own food are satisfied with any variety that produces fruit, and they have only a rudimentary understanding of heirloom seeds. I, too, was duped by the seed companies into thinking that hybrid is better. Hybrid tomatoes that we now find at the supermarket were bred to be tough enough to ride cross-country and still look like a tomato upon arrival. Unfortunately, it never did taste like a tomato!

True, heirloom tomatoes may not ship long distances, but they make up for it in taste, productivity, disease resistance, and other assets. Some of these seeds have been handed down from generation to generation for over one hundred years. What is so meritorious about a variety that would cause it to survive for so long? How exciting!

Too many of us took heirloom seeds for granted. We have either lost the seed altogether or have lost access to it. The need to educate our public about this important aspect of our heritage can't be overstated. It will be difficult but not impossible. We can start by saving and sharing.

Bill Best showing great-grandchildren Greta and Will Hess how to remove seeds from a candy roaster

An Offering of Thanks

Plant scientists sometimes refer to the Southern Appalachians as the "Vegetative Wonder of the World." This description usually refers to the native wild plants of the region, but it is true of the cultivated plants as well and continues into the modern day because of the Southern Appalachian people. We can all be thankful for the work of those individuals just discussed and of the countless numbers who, though they may not be as much in the public eye, save their seeds each year to give to the next generation of gardeners in their families and communities.

Since the 1940s, I have come to know hundreds of farmers and gardeners who love what they do. Some are highly educated with many degrees and important public responsibilities, while others have never attended school and cannot read and write. But they have in common a desire to maintain their plants, save their seeds, and pass these varieties on to the next generation. This attitude is best expressed in a song written by Jason Davis and Bryan Turner of Rose Hill, Virginia (reprinted with permission of the authors):

Save a Little Seed

I used to follow Papaw to that garden in his yard.
Beans were climbing up the cornstalks, which made the
picking hard
'Cause I was only 8 and the plants were 6 feet high.
Dodging bees and scratchy leaves I'd reach up to the sky.

He'd say, "We don't want this kind to disappear.
We need to dry a few to grow again next year."

Don't forget, don't throw away
Your ties to yesterday.
Look ahead, live your life.
Just remember those behind you
And always try to save a little seed.

He said, "My daddy grew those beans like his daddy did
before.
Best thing we had to eat those years during the war.
You can't buy them anywhere but they always produce.
I believe in keeping up with things that you can use.

"They show how our family's survived
And I'm more than proud to keep that heritage alive."

Don't forget, don't throw away
Your ties to yesterday.
Look ahead, live your life.

Just remember those behind you
And always try to save a little seed.

I think about Papaw when September rolls around.
My son follows me to the garden and we pull the bean vines
down.
I take a few pods, hull 'em in his little hands,
Look into his eyes, knowing he'll soon be a man.

I've found I have to stop and listen twice;
It seems like Papaw's talking, 'cause I'm giving his advice:

Don't forget, don't throw away
Your ties to yesterday.
Look ahead, live your life.
Just remember those behind you
And always try to save a little seed.

Words and music by Jason Davis and Bryan Turner

247 Hennegar Town Rd., Rose Hill, Virginia 24281
276-445-5440~betweenthetreesmusic@gmail.com

Judith Denny Whitehead Patteson, one hundred years old, holding some of Ardelia's Speckled Butter Beans, which she has grown for many years in Amherst, Virginia. "They're real reliable," she says. "I pick them while the pods are green and cook them like butter beans or lima beans with butter, water, and salt." *Photo by Denny Patteson*

For Further Reading

The following list is just a small sampling of the many books available for further reading on heirloom seeds, heritage fruit, and sustainable agriculture in general.

Buchanan, David. *Taste, Memory: Forgotten Foods, Lost Memories, and Why They Matter.* Foreword by Gary Nabhan. White River Jct., VT: Chelsea Green, 2012.

Calhoun, Creighton Lee, Jr. *Old Southern Apples: A Comprehensive History and Description of Varieties for Collectors, Growers, and Fruit Enthusiasts.* Revised and expanded edition. White River Jct., VT: Chelsea Green, 2011.

Estabrook, Barry. *Tomatoland: How Modern Industrial Agriculture Destroyed Our Most Alluring Fruit.* Kansas City, MO: Andrews McMeel, 2011.

Gift, Nancy. *A Weed by Any Other Name: The Virtues of a Messy Lawn, or Learning to Love the Plants We Don't Plant.* Boston: Beacon, 2009.

Kingsolver, Barbara, with Camille Kingsolver and Steven L. Hopp. *Animal, Vegetable, Miracle: A Year of Food Life.* New York: Harper Collins, 2007.

Lundy, Ronni, ed. *Cornbread Nation 3: Foods of the Mountain South.* Cornbread Nation: Best of Southern Food Writing. Southern Foodways Alliance, general editor John T. Edge. Chapel Hill: University of North Carolina Press, 2005.

———. *In Praise of Tomatoes: Tasty Recipes, Garden Secrets, Legends, and Lore.* New York: Lark Books, 2004.

Male, Carolyn J. *100 Heirloom Tomatoes for the American Garden.* New York: Workman, 1999.

Nabham, Gary Paul, ed. *Renewing America's Food Traditions: Saving and Savoring the Continent's Most Endangered Foods.* Foreword by Deborah Madison. White River Jct., VT: Chelsea Green, 2008.

Pellegrini, Georgia. *Food Heroes: 16 Culinary Artisans Preserving Tradition.* New York: Stewart, Tabori, & Chang, 2010.

Pollan, Michael. *The Omnivore's Dilemma: A Natural History of Four Meals.* New York: Penguin, 2006.

Ray, Janisse. *The Seed Underground: A Growing Revolution to Save Food.* White River Jct., VT: Chelsea Green, 2012.

Robe-Terry, Anna Lee. *Bootstraps and Biscuits: 300 Wonderful Wild Food Recipes from the Hills of West Virginia.* Parsons, WV: McClain, 1997.

Still, James. *Jack and the Wonder Beans.* Lexington: University Press of Kentucky, 1996.

Veteto, James R., Gary Paul Nabham, Regina Fitzsimmons, Kanin Routson, and DeJa Walker. *Place-Based Foods of Appalachia: From Rarity to Community Restoration and Market Recovery.* Published by Renewing America's Food Traditions (RAFT), www.raftalliance.org, 2011.

Index

Page references in italics denote illustrations